Instrument Science and Technology

Instrument Science and Technology

VOLUME 2

Edited by Barry E Jones

Department of Instrumentation and Analytical Science
University of Manchester Institute of Science and Technology

Adam Hilger Ltd, Bristol

Copyright © 1983 Adam Hilger Ltd

British Library Cataloguing in Publication Data
    Jones, Barry E.
    Instrument science and technology.
    Vol. 2
    1. Measuring instruments
    I. Title
    620'.0044   QC100.5
    ISBN 0-85274-753-5

Consultant Editor: **A E Bailey**, formerly of the National Physical Laboratory, Teddington, UK.

First published as a series of articles in *Journal of Physics E: Scientific Instruments* 1981/2, by The Institute of Physics.
This edition published by Adam Hilger Ltd, Techno House, Redcliffe Way, Bristol BS1 6NX, UK.
The Adam Hilger book-publishing imprint is owned by The Institute of Physics.

Printed in Great Britain by J W Arrowsmith Ltd, Bristol.

Preface

Measurement and instrumentation is an exciting growth subject covering many disciplines (Jones 1983a). There is strong evidence that implementation of automation in process and manufacturing industries has been seriously restricted by the lack of suitable instrumentation and analytical equipment, particularly at the plant-to-instrument interface. There are requirements for new measurement and control methods and devices for robotic, energy conservation, environmental control and biomedical applications. A recognition of these problems has resulted in a whole range of techniques and technologies being investigated with a view to producing novel and enhanced-performance sensing devices. It is evident that microelectronics, solid-state electronics, optoelectronics, fibre optics, ultrasonics, nucleonics, thin-films and modern methods of signal processing will play a significant role in these developments.

The science and technology of measuring instruments is now recognised to be a systemised discipline in its own right (Jones 1983b). Instrument designers and measurement scientists have sought to establish general principles and concepts of measurement and to present instrumentation and analytical equipment within logical frameworks. In recent years leading international experts in the field have produced relevant publications (for example Sydenham 1982) and more centres of excellence in the subject have been created in establishments of higher education (Jones 1983b).

In 1982 Adam Hilger Ltd published the first volume in an important new series of books on instrument science and technology (Jones 1982). The first volume comprises 13 articles written on topics which seem to be fundamental to all scientific instruments and which were originally published in the *Journal of Physics E: Scientific Instruments*. An introductory chapter about basic principles is followed by chapters on mathematical modelling, parameter estimation methods, dynamics, systematic design, feedback, reliability and ergonomics. The final five chapters relate to the fundamentals of noise and measurement.

This second volume in the series comprises 13 articles written with an emphasis on instrument technology and which were originally published in the *Journal of Physics E: Scientific Instruments* during 1981 and 1982. An introductory chapter about measurement errors and instrument inaccuracies is followed by chapters on kinematic design of fine mechanisms in instruments, displacement transducers based on reactive sensors, silicon micro-transducers, digital transducers, digital signal conditioning and conversion and advances in lock-in amplifiers. A chapter on correlation in instruments concentrates on cross correlation flowmeters, while a chapter on nucleonic instrumentation is concerned with the measurement of physical parameters by means of ionising radiation. The final four chapters are of a more general character, discussing measurement for and by pattern recognition, creative instrument design, future instrumentation and its effect on production and research, and the literature of instrument science and technology.

The 27 authors in the two volumes are leading experts in the field of measurement and instrumentation and come from four countries, Australia, East Germany, The Netherlands and the United Kingdom. Reflecting the educative role of the series, 22 of the authors are (or were) based in institutions of higher education, and of these only four authors come from departments of physics, with most being based in engineering or science/technology-based departments. Clearly technologists and engineers seem at present to be the main driving force behind the development of measuring instruments and the study of the general principles of measurement. At a first glance this situation seems rather surprising because of the multidisciplinary nature of the subject and the notion that study in the physical and natural sciences provides a broad basic training in experimental methods. One might expect more physicists (and chemists), particularly those in institutions of higher education, to see measuring instruments as tools of knowledge (Sydenham 1979) and to do some fundamental and original thinking about instrument science and technology.

v

Maybe the thrust of narrow specialism, excessive concentration on theoretical studies, 'big science' and defence projects and isolation from industry have had a debilitating influence on a substantial number of scientists.

Perhaps the search solely for originality and novelty has become an obsession and what we need now is a better understanding of knowledge that already exists and wider dispersion and a more profitable use of this knowledge. The *Journal of Physics E* series and these volumes are an attempt to move in this direction in the field of instrument science and technology.

Barry E Jones

References

Jones B E (editor) 1982 *Instrument Science and Technology* volume 1 (Bristol: Adam Hilger)
—— 1983a New transducers using microcomputers, fibre optics and thin films *Phys. Technol.* **14** 4–9
—— 1983b Instrument science and technology: a systemised discipline *Meas. Control* **16** (2) 63–71
Sydenham P H 1979 *Measuring Instruments: Tools of Knowledge and Control* (Stevenage: Peter Peregrinus)
—— (editor) 1982 *Handbook of Measurement Science Volume 1: Theoretical Fundamentals* (Chichester: John Wiley)

Present Addresses of Authors

M S Beck
Department of Instrumentation and Analytical Science, University of Manchester Institute of Science and Technology, Manchester M60 1QD, UK

M J Cunningham
Electrical Engineering Department, University of Manchester, Manchester M13 9PL, UK

J E Furse
6 Coleshill Road, Teddington, Middlesex TW11 0LJ, UK

A J Hugill
The School of Earth Sciences, The Flinders University of South Australia, Bedford Park, SA 5042, Australia

M L Meade
Faculty of Technology, The Open University, Walton Hall, Milton Keynes MK7 6AA, UK

S Middelhoek
Department of Electrical Engineering, Delft University of Technology, Delft, The Netherlands

B E Noltingk
Windwhistle, Nutcombe Lane, Dorking, Surrey RH4 3DZ, UK

D J W Noorlag
Department of Electrical Engineering, Delft University of Technology, Delft, The Netherlands

A R Owens
School of Electronic Engineering, University College of North Wales, Dean Street, Bangor, Gwynedd LL57 1UT, UK

R B J Palmer
Department of Physics, The City University, Northampton Square, London EC1V 0HB, UK

R Shaw
Industrial Division, SPL International Ltd, Battersea House, Battersea Road, Heaton Mersey, Stockport, Cheshire, UK

P H Sydenham
School of Electronic Engineering, South Australian Institute of Technology, Pooraka, South Australia

C J D M Verhagen
Nassaulaan 1, 2628 GA, Delft, The Netherlands

G A Woolvet
School of Mechanical, Aeronautical and Production Engineering, Kingston Polytechnic, Canbury Park Road, Kingston-upon-Thames, Surrey KT2 6LA, UK

Contents

Chapter 1

Measurement errors and instrument inaccuracies

M J Cunningham
Electrical Engineering Department, University of
Manchester, Manchester M13 9PL, UK

Abstract After a brief introduction to the activity of
measurement, the meaning of a measurement error is
discussed. The concept of systematic errors is described,
examples given and comments made on the difficulties of
ensuring that all significant systematic errors have been
considered. After an introduction to random errors, the
reasons are given why it is possible to state the range of
values within which the mean value of a quantity subject to
random variations will be from a small sample of
measurements and to give a probability that the statement is
true. The method of randomising systematic errors is given.
Various ways are described for stating the result of a
measurement where both systematic and random errors are
significant.

The statement of the performance of a measuring
instrument is reviewed and several proposals to unify the
method of statement by manufacturers described. The
advantages and disadvantages of these proposals to
manufacturers and users are discussed. Examples of the
specification of instrument performances are given. The
terms used are discussed and a method of obtaining the
likely operating error stated.

1 Measurement

The physical world is a world of change and apparent complexity. Heraclitus, the 5th century BC Greek philosopher,
said (Wheelwright 1959) 'You cannot step twice into the same
river, for other waters are continually flowing on'. It is
therefore always a surprise that the behaviour of much of this
physical world can be described by a very few simple formulations of law. Heraclitus also said (Wheelwright 1959) 'The
hidden harmony is better than the obvious'.

Many of these formulations of law are written in the language of mathematics.

Galileo said (Drake 1957) that science, 'is written in this
grand book, the universe, which stands continually open to our
gaze. But the book cannot be understood unless one first
learns to comprehend the language and read the letters in
which it is composed. It is written in the language of mathe--
matics and its characters are triangles, circles and other
geometric figures without which it is humanly impossible to
understand a single word of it; without these, one wanders
about in a dark labyrinth'.

The simplicity of the mathematical type formulations of law
is astonishing. For example, according to Coulomb's law, the
force of attraction between two electric charges is inversely
proportional to the square of their separation. It might be
that the exponent is not exactly 2 but just nearly 2. It can be
called $(2+q)$ where q is a small positive or negative number.
Recent measurements (Williams *et al* 1971) show that q is less
than 1×10^{-15}. The 'inverse square law' does appear to be an
inverse square law.

Once such simple formulations are available, two activities
are possible. The first is further investigation of the nature of
the physical world, an activity called science. The second is
that of effective application and design, an activity called
engineering. For example, large parts of the electrical and
electronics industries are based on the observations and
formulations of Faraday and their subsequent extension and
rendering in mathematical form by Maxwell.

Both of these activities, science and engineering, rely on
measurement. Measurement can be defined as the assignment
of a number to some quality of an object in terms of an
arbitrarily agreed unit value in such a way that the number
faithfully represents the properties of that quality of the object.

1

Measurement thus is a means of gaining access to the world of the formulation of law and the process of reasoning which accompanies it. Feynman (1965) says, in relation to the mathematical statement of law, 'mathematics is a language plus reasoning'. Measurements are therefore made in terms of the framework of the laws and made possible by the realisation and maintenance of a satisfactory system of units. The satisfactory establishment of the unit is in itself an interesting, if largely unnoticed, activity (Vigoureux 1971). None of the subsequent activities of measurement can take place without the existence of the one.

The meaning of the definition of measurement is clear enough for a quality directly observable by the senses. For example a metre rule can be used to ascribe a number to the quality length of an object in terms of the agreed unit, the metre.

What about qualities not observed by the senses directly, or only vaguely sensed, such as electric current? A device has to be found to make the quality observable. Such a device is called a measuring instrument. Although the human body is equipped with five powerful senses, sight is the sense used almost exclusively for measurement.

One of the main reasons for this is that the visual length is one of the very few stimuli (Schiffman 1976) that has a linear relation between perceived stimulus and the magnitude of the stimulus itself. This means that half and double the length of a line can be judged by eye in a very satisfactory manner. The same is not true of most other stimuli.

Examples of instruments which transform the quantity to be measured into length are very numerous and would include

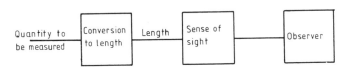

Figure 1   Measurement by conversion to length.

mercury-in-glass thermometers, barometers, deflection voltmeters, ammeters, cathode ray oscilloscopes, analogue clocks, pressure gauges and flow meters. An enormous number of measurements can be illustrated by figure 1.

Another vast range of measurements is made by transforming the quantity to be measured into discrete things and counting these, usually over a specified period of time. Examples of this are digital voltmeters, digital multimeters and quartz thermometers. These can be represented diagramatically by figure 2.

Other implementations of measurement can be represented in a similar way.

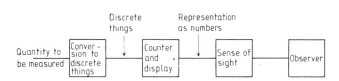

Figure 2   Measurement by conversion to counting.

2   Measurement errors

2.1 *Introduction*

Since the quantity to be measured, the instrument by which the quantity is brought to the awareness of the observer via the senses and the state of the observer are in continual change, what is the significance of the result of a measurement? From one point of view every measurement is right since it gives rise to the number ascribed to that quantity by that instrument by that observer at that instant. This approach is not fruitful however because the aim of the measurement is to relate it to other measurements by other instruments and other observers at other times.

An International Organisation of Legal Metrology definition of the error of a measurement (OIML 1968) is the discrepancy between the result of the measurement and the true value of the quantity measured.

Since, to modify the first quotation of Heraclitus, 'You cannot make the same measurement twice', even with a perfect instrument and observer, the result of a measurement would never be exactly the same twice, except for the trivial example of counting objects. The instrument and observer also introduce errors.

The result of the measurement is what is available, but the 'true value' of the quantity, free from any error, is what should be communicated. The 'true value' is unknowable but it is necessary to state from the result of measurement within what range of values the 'true value' is likely to occur and the confidence that can be placed on the validity of this statement. This procedure is cumbersome but inevitable. The statement of the result of a measurement without any indication of the range of values within which, with a certain confidence, the 'true value' can be assumed to be is almost worthless.

Initially the way to state the results of a measurement will be discussed assuming errors owing to perturbations of the quantity to be measured, the instrument and the observer. In §4 the special case of specifying the performance of an instrument will be considered.

2.2 *Systematic errors*

It is convenient initially to divide errors into the two classes of systematic and random, although, as will be shown later, the distinction is not always as distinct as might at first appear.

OIML define systematic error as (OIML 1968) 'an error which, in the course of a number of measurements, made under the same conditions, of the same value of a given quantity, and which either remains constant in absolute value and sign, or varies according to a definite law when the conditions change'. The examples given for a systematic error are as follows:

'(a) The error which results from a weighing by means of a weight whose mass is taken to be equal to its nominal mass of 1 kg whereas its conventional true mass is 1.010 kg.

(b) The error which results from using, at an ambient temperature of 20 °C, a rule gauged at 0 °C without introducing a suitable correction.

(c) The error which results from the use of a thermoelectric thermometer whose circuit suffers from parasitic thermoelectric effects.'

There will be in general a large number of effects which can change the result of the measurement in this way. A great deal of intelligence and experience is required to decide which are the significant effects. Some components of the systematic error are revealed by performing experiments in which the conditions of the measurement are changed in an orderly way and noting the changes in the result of the measurement. It is, however, always possible that a significant systematic error will remain unsuspected. This seems to be common and frequently the results of a measurement are stated with extreme optimism.

Since a systematic error is simply an error much care must be taken in designing the measurement to reduce the size of systematic errors as much as possible and then to try to discover the size of the remaining significant systematic errors. It becomes increasingly difficult, as the tolerable size of the error becomes smaller, to say with complete confidence that there are no systematic errors of significant size that have not been considered.

Sometimes it is possible to measure the quantity using quite a different principle, which will have mostly different systematic errors, to reveal an unsuspected systematic error in the original measurement or to increase the confidence in the statement of the maximum systematic error.

As an example of this at the highest level, the various methods of realising the ampere are at present under the suspicion of suffering from unknown systematic errors (Taylor 1976). In practice, the ampere is maintained at standard laboratories such as the National Physical Laboratory in the United Kingdom by means of stable voltage sources and standard resistors. The ratio, K, between this maintained unit of current and the SI ampere can then be established in one of two basic ways: either directly, by using some form of current-force balance, or indirectly by the appropriate combination of the values of various fundamental constants.

There are three main indirect methods:
 (i) the gyromagnetic ratio of the proton by both low and high field methods,
 (ii) the Faraday,
 (iii) the Avogadro constant.
The values of K found by these three methods not only differ among themselves, but also disagree with those given by the direct current balance method to such a degree that unsuspected systematic errors may be present (Petley 1979).

The General Conference on Weights and Measures (CGPM 1979) recently recommended because of these discrepancies 'the continuation and intensifying of research on both the direct realisation of electrical units and the indirect realisation through the determination of physical constants'.

Systematic errors can be significant at any level of measurement and the user of measuring instruments, particularly digital ones, needs to develop the frame of mind whereby he does not easily believe the numbers presented to him to be acceptably close to the true value of the quantity he would like to measure.

2.3 Random errors

If a quantity is measured with an instrument with good discrimination, sometimes called resolution, it is often found that repeated readings are not identical, as would be the case with only constant systematic errors. In such a case it is interesting to consider how best to convey the result of the measurement.

In order to indicate the result of a series of measurements of the 'same' quantity to others, all the results can be presented in tabular form. Rather more easy for the mind to work with is a graphical presentation. Vigoureux (1966) has suggested a

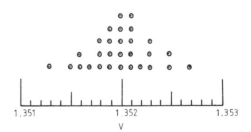

Figure 3 Vigoureux's method for displaying results.

simple method in which the results are entered on a scale as they are taken. Whenever a value is repeated, the point is placed on the line above the previous line of that value. This is illustrated in figure 3 and this method was used by Vigoureux to present the results of his determination of the ampere (Vigoureux 1965). All the results and their scatter are clearly

3

presented. This display is really a particular form of a graphical presentation of scattered values called a histogram.

2.4 *The histogram*
When large numbers of measurements are made of the 'same' quantity, it is convenient to represent the results as a frequency distribution, a particularly convenient form of which is the histogram shown in figure 4. The area of each rectangle is proportional to the number of results within the limits given by the edges of that rectangle. If the widths of the rectangles are made equal, that is difference between the limits of the

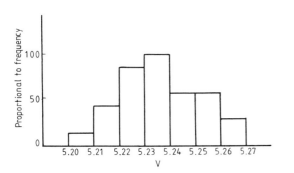

Figure 4 The histogram.

classes are equal, then the heights of the rectangles are proportional to the frequency of the results of measurements between the limits shown.

2.5 *The limiting mean of the results of measurements*
Although the histogram displays much of the information of the results of the measurements, it is often useful to describe the series of results by numbers. One such number is the mean of the series of results.

Dorsey (1944) said 'The mean of a family of measurements – of a number of measurements for a given quantity carried out by the same apparatus, procedure and observer – approaches a definite value as the number of measurements is increased. Otherwise, they could not properly be called measurements of a given quantity. In the theory of errors, this limiting mean is frequently called the true value, although it bears no necessary relation to the true quaesitum, to the actual value of the quantity that the observer desires to measure. This has often confused the unwary. Let us call it the limiting mean'. Eisenhart (1963) has called this the 'Postulate of Measurement'.

2.6 *The standard deviation*
Following from this postulate it is possible to obtain a number which gives an indication of the scatter of the results of the measurements. There are several ways of doing this.

The most commonly used is called the standard deviation, which is defined as the value approached by the square root of the average of the sum of the squares of the deviations of individual measurements from the limiting mean as the number of measurements is indefinitely increased.

In symbols

$$\left(\frac{1}{n}\sum_{i=1}^{n}(x_i-m)^2\right)^{1/2}\to\sigma$$

as

$$n\to\infty$$

where σ is the standard deviation, m the limiting mean and x_i the ith measurement. Variance is defined to be the square of the standard deviation σ^2.

This definition of standard deviation as the root mean square deviation from the limiting mean is an acceptable way of specifying the scatter of the results of measurement, although the word 'standard' does seem to give this definition perhaps an unwarranted appearance of fundamentality. The term probable error, defined to be the deviation from the limiting mean which is just as likely to be exceeded or not, although acceptable in principle, has an unfortunate name. Moroney (1951) gives the comment that 'it is neither an error nor probable'. This term will not be used in the remainder of this article.

The limiting mean and the standard deviation can be given for any very large number of measurements of the 'same' quantity.

2.7 *Samples of results of measurements*
Of course, usually only a few repeated measurements of the 'same' quantity are made and so it might appear that the concepts of limiting mean and standard deviation are of little practical use. Perhaps surprisingly this is not the case.

A theorem exists called the Central Limit Theory. Mood (1950) describes this theorem as the 'most important theorem in statistics from both the theoretical and applied points of view'.

The theorem states:
'If a population has a finite variance σ^2 and mean m, then the distribution of the sample mean approaches the normal distribution with variance σ^2/n and mean m as the sample size n increases.'

Nothing is said in the theorem about the form of the population distribution function, which is just as well since it cannot be known from a few measurements. The theorem does state that the distribution of sample means of independent measurements will be approximately normal about the population limiting mean with a standard deviation for this distribution of $\sigma/n^{1/2}$. $\sigma/n^{1/2}$ is called the standard error of the mean.

For a series of measurements with a more or less symmetrical histogram and small deviations from the mean compared with the magnitude of the mean, the distribution of sample means is very close to normal even for n as small as 3 or 4 (Ku 1969). These conditions apply for a large number of measurements.

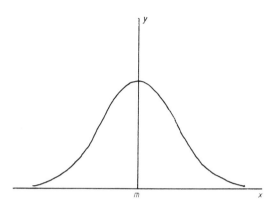

Figure 5 The normal distribution.

2.8 *The normal distribution*

The normal distribution referred to in the Central Limit Theorem is well known in statistics. It was first given by De Moivre (1733) while living in England as a consultant on gaming and insurance. The distribution is illustrated in figure 5.

Mathematically the normal distribution is given by

$$y = \frac{1}{\sigma(2\pi)^{1/2}} \exp\left[-(x-m)^2/2\sigma^2\right].$$

The normal distribution is one for which the relative frequency of the data having a value between x and $x + \delta x$ is $y\delta x$, where y is given by this equation. y is therefore called the probability density of the distribution.

The area under the curve between two x values is therefore the probability that the quantity will occur between these two values. The total area under the normal distribution is unity.

2.9 *Statement of the population mean from a sample of measurements*

The Central Limit Theorem says that to a good approximation, the means of fairly small samples will lie on a normal distribution with mean equal to the population mean and standard deviation equal to $\sigma/n^{1/2}$ where σ is the population standard deviation.

The Central Limit Theorem is powerful because it enables statements to be made on the probability that the population mean is less than a specified deviation from the sample mean.

Table 1 gives the probability that a normally distributed quantity will be within r standard deviations of the mean.

Table 1

r	% probability
0.6745	50
1	68.27
2	95.43
3	99.73
∞	100

So if the sample mean is m_s and the population mean, the quantity we are seeking, is m, then in more than 95% of samples, the population mean will be within two standard deviations of the sample distribution.

So

$$-2\sigma/n^{1/2} < m_s - m < +2\sigma/n^{1/2}$$

or

$$m_s - 2\sigma/n^{1/2} < m < m_s + 2\sigma/n^{1/2}$$

will be correct for over 95% of a large number of sets of samples.

This statement would allow statements to be made, from a single small sample of measurements, on the probability of the population mean being more than say $2\sigma/n^{1/2}$ away from the sample mean. Unfortunately, it is necessary to know σ, which is the standard deviation of the whole population and is found by a very large number of measurements!

All is not lost, however, because the population standard deviation σ can be estimated from the sample standard deviation, σ_s. Of course, σ_s underestimates the population standard deviation and so the probability statement has to be re-written that

$$m_s - t\sigma_s/n^{1/2} < m < m_s + t\sigma_s/n^{1/2}$$

is true for a certain percentage of samples. The multiplier t depends on n and the probability level chosen and is called Student's t.

The distribution of t was obtained mathematically by Gosset (1908) and published under the pen name of 'Student'.

Values of t have been evaluated for the normal distribution and are given in table 2.

To find t, it is necessary to know the number of degrees of freedom of the calculated sample standard deviation σ_s. Since m_s is found from the same n results, the last value is fixed by m_s and the other $(n-1)$ values. The degree of freedom

Table 2 Values of Student's *t*

Degrees of freedom	$P = 68.3\%$ (1σ)	$P = 95\%$	$P = 99\%$	$P = 99.73\%$ (3σ)
1	1.8	12.7	64	235
2	1.32	4.30	9.9	19.2
3	1.20	3.18	5.8	9.2
4	1.15	2.78	4.6	6.6
5	1.11	2.57	4.0	5.5
6	1.09	2.45	3.7	4.9
7	1.08	2.37	3.5	4.5
8	1.07	2.31	3.4	4.3
9	1.06	2.26	3.2	4.1
10	1.05	2.23	3.2	4.0
15	1.03	2.13	3.0	3.6
20	1.03	2.09	2.8	3.4
30	1.02	2.04	2.8	3.3
50	1.01	2.01	2.7	3.2
100	1.00	1.98	2.6	3.1
∞	1.00	1.96	2.58	3.0

is therefore $n-1$. Strictly the estimated standard deviation is given by

$$\left(\frac{1}{n-1}\sum_{i=1}^{n}(x_i-m)^2\right)^{1/2}.$$

However, except for very small n, the change is small (Müller 1979).

For example, if a series of five measurements of current give a mean of 1.123 A and a standard deviation of 0.005 A, then for a 99% probability level and degree of freedom 4, the table shows that $t = 4.6$.

The statement can therefore be made that there is a 99% probability that the population mean will be in the range

$$1.123 - \frac{4.6 \times 0.005}{5^{1/2}} < m < 1.123 + \frac{4.6 \times 0.005}{5^{1/2}}$$

or

$$1.113 < m < 1.133.$$

The ability to assign a probability that the population mean will be within a specified range of the mean of a fairly small sample is the basis of the application of statistics to measurement.

This above coverage of this very complicated subject is of necessity a simplified one. The standard text books on the subject, of which there are a vast number, give a more full treatment, for example Brownlee (1960), Mood (1950).

2.10 *Randomising systematic uncertainties*

It is not always clear whether an influencing quantity produces systematic or random errors. For example for a sample of measurements taken close together in time, humidity might well produce a systematic error, whereas if readings are taken for the sample over a long time scale humidity might well produce random errors.

This suggests one method that can be used to reduce the effect of or reveal systematic errors. By intelligent experimental technique it is possible to convert some systematic errors into random ones. At least their presence is then obvious!

3 The statement of errors

It will be clear that to be of maximum use, the statement of the result of a measurement or series of measurements should include a statement about the errors involved. For accurate measurements this should certainly be done very fully, however for routine low accuracy measurements this would be unnecessary. Nevertheless every statement should include enough information, often in the form of a sentence, which makes clear the intention of the author. This is necessary owing to the lack of agreed procedure in the statement of errors.

It is always possible to state the systematic and random errors separately. Campion *et al* (1973) have advocated this approach.

They recommend that the components of the random uncertainty should be listed in sufficient detail to make it clear whether they would remain constant if the experiment were repeated. The components of the systematic uncertainty should be listed, expressed as the estimated maximum value of that uncertainty. The method used to combine systematic uncertainties should be made clear. The authors do not recommend the combination of random and systematic uncertainties.

On the other hand Müller (1979) recommends random and systematic errors should have applied to them the general propagation law of errors. Each error is represented by the measured standard deviation, or the best estimate available, and these are combined using this law. Müller suggests this leads to 'a natural and unambiguous evaluation of the overall uncertainty to be associated with an experimentally determined quantity'.

This method, the author states, should be applicable to any 'level of metrology', thus unifying the different procedures hitherto used.

In another recent publication on the subject by Hayward (1977) there are recommendations that the total random uncertainty and the total systematic uncertainty should always be stated separately. If a value for the overall uncertainty is also required, then the total random uncertainty should be added to the total systematic uncertainty in quadrature.

Since there cannot be a theoretically justifiable way of combining random and systematic errors, it is not surprising that several methods have been proposed (Wagner 1980).

The International Bureau of Weights and Measures (BIPM) is actively seeking to find international concensus on the statement of the results of measurement (Giacomo 1979, BIPM 1980).

Even if a somewhat arbitrary method is used to combine random and systematic errors, most would agree that the components should also be given in full and the method of their combination stated.

The minimum acceptable statement of the result of a measurement would be:

'The current was 1.06 A. All errors are negligible to this degree of rounding.'

3.1 Rounding the statement of the result of a measurement

The number of decimal places to which the result of a measurement is stated in the absence of any other statement implies the level of errors expected. With decimalisation reaching consumer products in the United Kingdom the examples of poor rounding are numerous. For example, the information on a packet of seeds states that the plants will grow to 91.5 cm high. This implies the height will be between 91.45 cm and 91.55 cm high. Before decimalisation, the plants were stated to grow to 3 ft high, implying the height will be between 2 ft 6 in and 3 ft 6 in. It has to be assumed that a new super constant height plant has not been developed!

4 The specification of the errors of an instrument

When a potential user would like to select an instrument from those available from the manufacturers he would need to assess the errors introduced by the instrument in his particular measurements. Since the conditions under which the instrument is used will vary from user to user, the manufacturer faces an almost impossible task in making general statements about the error introduced by the instrument. There are clearly commercial pressures to state the error that might be expected in favourable circumstances. It is therefore a fairly lengthy task for the user to deduce from this the error to be expected in his own circumstances. Since manufacturers do not all state the errors and circumstances in the same way, the comparison of the expected performance of instruments from several manufacturers can be no light task.

The International Electrotechnical Commission (IEC) has over the last 10 years attempted several times to suggest ways in which the errors of an instrument could be stated. These will now be briefly described to illustrate the range of approaches that are possible in this difficult area.

IEC Publication 359 (1971) suggested a method for describing the performance of electronic measuring equipment. It was hoped that the adoption of this method by the industry would lead to easier and better comparison of equipment specifications by the user.

The publication specifies a large number of influence quantities. An influence quantity is defined to be any quantity, generally external to the apparatus, which may affect the performance of the apparatus. Influence quantities include climatic conditions such as ambient temperature, relative humidity of the air, velocity of the ambient air and so on, mechanical conditions such as operating position, ventilation, vibration, mechanical shock and so on, supply conditions such as mains supply voltage, frequency and so on and fields and radiations.

The publication also states ranges of many of these influence quantities. The manufacturer is then asked to state the error given by the most adverse combination of all the influence quantities.

This 'worst case' approach has much to commend it from the user's point of view. However since many instruments are significantly affected in performance by several influence quantities, the testing needed to verify the stated error is considerable. Also, a manufacturer quoting the inevitably large error would be at a commercial disadvantage.

By contrast IEC Publication 51 (1973), for example, suggests another approach. In this, errors, called intrinsic errors, are determined under a narrow range of influence conditions called reference conditions. These errors can be easily checked by most users. Some indication of the actual error to be expected in use is gained from stated 'variations'. Variation is the change in error produced by each influence quantity separately when changed from its reference value to an extreme of a stated operating range, while all the other influence quantities are kept at their reference values. There is therefore no indication of what might happen in actual use when several of the influence quantities would not be at their reference values.

The IEC has been considering the revision of IEC Publication 359 with the aim of combining the advantages and avoiding some of the disadvantages of the above two approaches. As a part of these considerations the IEC published a Draft Standard (1980). In this there is an attempt to give an indication of the actual error likely to occur while under operating conditions while on the other hand allowing the relatively easy checking of the stated errors.

In order to simplify, several rules were proposed which include the following. For most applications only three influence quantities cause significant error. The effects of these three influence quantities are independent. The effect on performance by an influence quantity is linear or a second order effect. A probable operating error is stated from the square root of the sum of the squares of the intrinsic error and the variations produced by the influence quantities.

It must be stressed that these last suggestions are only at a draft stage. A meeting of the IEC technical committee TC66 decided in September 1980 in the light of comments on the whole draft to rewrite it instead as a general document. This would primarily define regulations for the uniform specification of errors of all apparatus covered by the standard.

Nevertheless, the general thinking behind these latest proposals does offer some basis for a method of stating the performance of instruments acceptable to user and manufacturer alike. Any move towards the adoption of a common method of stating the performance of instruments would be of assistance to the user.

Hayward (1977) has given an interesting proposed standard procedure for measuring the repeatability and estimating the accuracy of industrial measuring instruments.

5 Some examples of instrument inaccuracy

There are a number of terms used to specify the performance of an instrument which are confusing to the mind. An example of such a term is the 'instrument accuracy'. The reasonable man might understand by this term the accuracy of the instrument. This turns out not to be the case. The quantity specified under this heading is part of the inaccuracy of the instrument and usually relates closely to the intrinsic error discussed in the previous section.

Examples of instrument specifications will now be given and the method for arriving at the likely operating error discussed.

5.1 A deflection multimeter

The following specification, although fictitious, is typical of this sort of instrument:

Accuracy (DC voltage and current) ±2% full scale deflection
        (AC voltage and current) (50 Hz) ±3% full scale deflection

Sensitivity DC voltage 20 000 Ω/V
         AC voltage 2 000 Ω/V

Frequency response < 3% variation from reading at 50 Hz
        (10 Hz–10 kHz)

Temperature effect 0.2%/°C.

The loading effect of the voltmeter, often referred to by the ambiguous name sensitivity, can give rise to a very large systematic error. Usually the desired result of the measurement is the value of the measured quantity without the measuring instrument present. For a voltmeter, the presence of the instrument can significantly change the potential difference to be measured. For example, consider the circuit shown in figure 6. Without the voltmeter connected the potential difference between points A and B would be 5.00 V. When the specified voltmeter in its 5 V range is connected to points A and B a 100 kΩ resistance will be introduced between A and B. The effective resistance between A and B would be 33 kΩ. The potential difference between A and B would therefore be 3.75 V. Even if the voltmeter had no other source of error, the percentage error of this measurement would be 25%. It is clear that this systematic error must be considered carefully whenever a measuring instrument is used. Since the voltmeter is the means of finding the value of the potential difference,

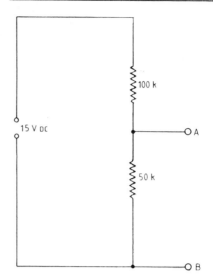

Figure 6 The loading effect.

obtaining the value in the absence of the voltmeter does raise some difficulties. Of course a higher resistance voltmeter can always be used. If this is not available, a similar voltmeter can be connected to the same points as the original voltmeter. If the reading of the original voltmeter changes, then the original voltmeter must have significantly affected the potential difference to be measured.

The frequency of an AC voltage can be regarded as an influence quantity when the amplitude of the AC voltage is required. When being used in this way for measurements other than at 50 Hz, a systematic error of 3% has to be assumed. More information from the manufacturer would be useful since it could be argued that the measurement would never be exactly at 50 Hz and so the 3% systematic error must always be assumed.

Another systematic error occurs owing to temperature changes. The calibration temperature is not stated, but presumably could be found by contacting the manufacturer. Then a systematic error of 0.2% for every degree celsius deviation from that temperature must be assumed.

The only remaining figure supplied is called the 'accuracy' of the instrument. In the terminology of the last section this is the intrinsic error. Although little of the random component of the result of a measurement is due to the instrument, such as there is, for example noise, will be included in this term. This term should also include the maximum error owing to all other influence quantities. That it does is often hard to believe. Such things as linearity, hysteresis, instrument attitude, electric and magnetic fields and temperature gradients in their worst combination are likely to contribute errors in excess of 2% of the full scale deflection. In the absence of further information, the user is forced to vary any unstated influence quantity suspected of introducing significant systematic errors in order to reveal the size of the error.

It is possible to combine the intrinsic error and all the variations produced by significant influence quantities. The method suggested in the last section to find an estimate for the operating error was by taking the square root of the sum of the squares of the intrinsic error and all the variations produced by influence quantities. Of course the error in a particular measurement can be larger than this estimate.

Clearly the instrument cannot be said to have an inaccuracy of 2% on DC ranges. The error in an operating condition can be much larger than this. The error of the instrument is not a constant amount. The user should ask himself 'What is the error of the instrument now?'

5.2 *A digital frequency counter*

The following are the specifications of a fictitious, but typical, frequency counter:

Input impedance	1 MΩ in parallel with 25 pF
Temperature stability	± 4 parts in 10^6 over 20 °C to 40 °C
Aging rate for crystal	± 2 parts in 10^6 per month, 2 months after delivery.

The loading effect, although not as serious as for the deflection multimeter, can still give rise to problems by changing the frequency to be measured on connecting the instrument, for example, by changing the characteristics of an oscillator. This effect is most likely to produce an error at high frequency since the input impedance of the instrument is much lower in this case than for low frequency measurements.

Over the specified temperature range, the contribution to the error will be less than the stated amount. The statement is rather pessimistic for normal environments and a smaller error contribution could presumably be obtained for a small temperature range.

The aging error gives some indication of the error from this source. With regular calibration, this error can be made very small.

The only other error is the normal ± 1 error of a digital instrument. This can be made small by including a sufficiently large number of counts.

It has to be assumed that all other errors, such as random errors, are negligible. The comments on obtaining the operating error also apply to this instrument.

6 Conclusions

Except for the trivial case of counting objects, it is not possible to give exactly the value of a quantity from the result of a measurement of it. When the result of a measurement is stated the aim must be to communicate as much as possible and this means a statement of the probability that the true value lies within a stated range. This statement would be very full for high accuracy measurement but could be as simple as

'the errors are negligible to this degree of rounding'.

The statement of the performance of an instrument is a subject with many technical and commercial considerations. There are recent proposed procedures for specifying the performance of an instrument which appear to overcome some of the objections to previous proposals. Any move towards a common method of statement of instrument performance would be welcomed by the users.

References

BIPM 1980 *Rapport BIPM 80/3*

Brownlee K A 1960 *Statistical Theory and Methodology in Science and Engineering* (New York: Wiley)

Campion P J, Burns J E and Williams A 1973 *A Code of Practice for the Detailed Statement of Accuracy* (London: HMSO)

CGPM 1979 The Sixteenth Conference was reported by Giacomo P 1980 News from the BIPM *Metrologia* **16** 55–61

De Moivre A 1733 Approximatio ad Summam Terminorum Binomii $(a+b)^n$ in Serien Expansi *Miscellenea Analytica de Seriebus et Quadraturis* (Second Supplement)

Dorsey N E 1944 The velocity of light *Trans. Am. Phil. Soc.* **34** 1–110

Drake S 1957 *Discoveries and Opinions of Galileo* (New York: Doubleday)

Eisenhart C 1963 Realistic evaluation of the precision and accuracy of instrument calibration systems *J. Res. Nat. Bureau of Standards* **67C** 161–87

Feynman R 1965 *The Character of Physical Law* (London: BBC)

Giacomo P 1979 News from the BIPM *Metrologia* **15** 51–4

Gosset W S 1908 The probable error of a mean *Biometrika* **6** 1–25

Hayward A T J 1977 *Repeatability and Accuracy* (London: Mechanical Engineering Publications)

IEC 1971 *Publication 359 Expression of the Functional Performance of Electronic Measuring Equipment* This document was the basis for the British Standards Institution publication BS 4889 (1973) British Standard method for specifying the performance of electronic measuring equipment

IEC 1973 *Publication 51 Specification for Direct Acting Indicating Electrical Measuring Instruments and their accessories* This document was produced in identical form as the British Standard BS 89 (1977)

IEC Draft Standard 1980 *Draft Standard for the Expression of the Performance of Electrical/Electronic Equipment for Measurement, Control or Analysis* This document was issued by the British Standards Institution as Draft for Public Comment 80/250 77 DC

Ku H H 1969 Statistical concepts in metrology *National Bureau of Standards Special Publication 300* **1** 296–330

Mood A M 1950 *Introduction to the Theory of Statistics* (New York: McGraw-Hill)

Moroney M J 1951 *Facts from Figures* (London: Penguin)

Müller J W 1979 Some second thoughts on error statements *Nucl. Instrum. Meth.* **163** 241–51

OIML 1968 *Vocabulary of Legal Metrology* The British Standards Institution issued an unofficial English translation in PD 6461

Petley B W 1979 The ampere, the kilogram and the fundamental constants – is there a weighing problem? *NPL Report QU52*

Schiffman H R 1976 *Sensation and Perception* (New York: John Wiley)

Taylor B N 1976 Is the present realization of the ampere in error? *Metrologia* **12** 81–3

Vigoureux P 1965 A determination of the ampere *Metrologia* **1** 3–7

Vigoureux P 1966 Errors of observation and systematic errors *Contemporary Physics* **7** 350–7

Vigoureux P 1971 *Units and Standards for Electromagnetism* (London: Wykeham Publications)

Wagner S R 1980 Combination of Systematic and Random Uncertainties *Conference on Precision Electromagnetic Measurement* (New York: IEEE)

Wheelwright P E 1959 *Heraclitus* (Princeton: Princeton University Press)

Williams E R, Faller J E and Hill H A 1971 New experimental test of Coulomb's law; a laboratory upper limit on the photon rest mass *Phys. Rev. Lett.* **26** 721–4

Chapter 2

Kinematic design of fine mechanisms in instruments

J E Furse

Engineering Services, National Physical Laboratory, Teddington, Middlesex, UK

Abstract This article considers some of the basic principles involved in kinematics or the theory of mechanical location and movement which involves no redundant locations or restraints on the mechanical component parts of an instrument or unit.

Kinematic location and the various degrees of freedom are described with reference to stationary, sliding, rotating and screw couplings. Reference is also made to some of the associated problems involved such as the lobing of balls and the deflection of members.

1 Introduction

In Science and Industry accurate measurement is of the utmost importance. This is particularly so in engineering and, though it is essential that the design of instruments for this purpose should keep pace with the rapid advances in scientific knowledge and modern industrial processes, there are some basic principles which, if followed will produce sound and relatively simple designs which require a minimum of highly skilled work to produce them.

This is regardless of the type of instrument, be it for the measurement of length, volume, area, mass, force, temperature or pressure since, in general, the problem to be overcome lies in the magnification of small movements. This can be achieved by one or more of several ways, either mechanically, electrically, electro-mechanically, optically or by a combination of optical and mechanical methods, electronically or pneumatically.

In recent years the use of pneumatic devices has waned whilst the use of optical methods, particularly that using a laser light source, has grown in popularity. This has many advantages over a purely mechanical method since a much higher degree of magnification can be attained with often only the mounting of the mirror or a cube corner component requiring special mechanical consideration.

The electromechanical transducer also has considerable popularity. In this case an electrical out of balance, due to the movement of a measuring plunger connected to a floating armature housed within coils, which forms part of a bridge circuit, causes a current to flow to be detected by a meter or other means of observation.

Thus, in general, the operation of a precision measuring instrument is that of the amplification of small initial movements to movements of a larger magnitude, which may be observed visually on a line or scale.

2 Definitions

It is appropriate at this stage to refer briefly to the terms frequently used to simplify the conception of mechanisms, namely elements, links and lower and higher pairs.

An instrument is made up of a number of moving parts connected together so as to be capable of relative motion and the two parts that are in contact, and between which there is relative motion, are known as a 'pair' or 'coupling' and the two parts forming the pair or couple are referred to as 'elements'.

Relative motions commonly encountered are sliding, rotating and screwing. These come under the heading of 'lower pairs'.

All other pairs are a combination of sliding and rotating and are known as 'higher pairs'.

A connection between any two pairs is termed a 'link'.

For an instrument to be successful it must be precise, accurate, sensitive, have consistent repeatability in its readings and any inherent hysteresis must be small.

Considering these features individually, it is possible to make a precise measurement but this is no guarantee that it is accurate but an accurate measurement must be made with precision.

Accuracy is of the greatest importance and it may be stated that an instrument is accurate when its indication agrees with the true value of the quantity being measured. Factors affecting the accuracy of operation of the instrument may be wear, temperature, atmospheric pressure, humidity and the method of use and manipulation of the instrument.

Sensitivity, again of great importance, may be defined as the ratio of the displacement of the indicating element to the measured quantity. The sensitivity of an instrument should be at least ten times the accuracy of determination required for the measured quantity. In instruments where no physical contact is made, as in interferometers or pneumatic devices, the definition above can be confusing. It can be expressed as the smallest change in the value of the measured quantity which produces a perceptible response in the indicator of the instrument. This should not be confused with sluggishness or lack of response caused by friction and backlash in the mechanism.

Repeatability is defined as the range of variation in the reading shown by the instrument over repeated measurements during which the conditions remain unchanged. Lack of repeatability of reading is mostly caused by the small clearances necessary in some couplings to permit relative movement of the elements, and in some cases by the introduction of lubrication.

Hysteresis is also a cause of lack of repeatability in a mechanism. This is indicated when an instrument is operated over its full range of movement in one direction, with the reading noted at intervals against standards, then, using the same standards, operating the instrument in reverse back to the initial value. It will be found that two curves are produced, which in general form a loop as in figure 1. The boundaries of the hysteresis loop will represent the extreme variations of the range of uncertainty in readings taken with the instrument when all causes, except those inherent in the instrument, are eliminated.

3 Principles of alignment

Factors which contribute to a successful design of a measuring instrument and which have been mentioned above namely accuracy, sensitivity, repeatability and hysteresis can only be determined fully by calibrating a completely finished instrument.

Therefore, in order to ensure that an instrument can be produced in which errors in each of these factors are within acceptable limits, it is essential that certain principles are observed from the earliest stages of the design.

One of the most important is the principle of alignment

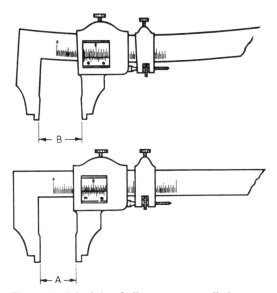

Figure 2 Principle of alignment as applied to a vernier.

which applies chiefly to guideways and slides since any errors present in the ways are generally reflected in any measurements made.

An obvious example of this is in the popular vernier calliper. This is not an instrument of high precision, partly because of its low magnification but mainly because of the principle underlying its design. Figure 2 shows that if the beam of the calliper is not straight the size of any part measured at the jaws B is not truly indicated by the movement of the slide along the scale.

The same conditions of alignment apply to machines of many types. A particular important example is the jig borer in which it is inevitable that the scales, lead screw or other

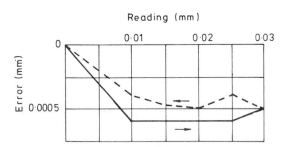

Figure 1 Hysteresis loop of dial indicator.

methods of measurement be at an appreciable, and in fact varying, distance from the plane of the work. Errors from this cause will be annulled if the axis of measurement is made co-linear with the axis of motion of the machine slide.

This principle of alignment should be maintained, if it is at all possible, in any measuring device. However, in machines such as the jig borer a compromise is usually necessary from other considerations.

An ordinary screw micrometer accords with this principle which is a primary factor in the maintenance of the accuracy of such instruments over long periods of hard usage. Similarly where an accurate screw thread is used for translating motion to a machine or other slide the axis of the screw thread should be in the plane of measurement.

4 Kinematic principles

Next in importance to the alignment principle in the design of couplings and pairs is the observance of the kinematic principle, as originally recognised by Clerk Maxwell in 1876 who was one of the pioneers in this field.

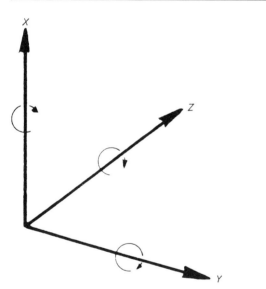

Figure 3 The six degrees of freedom.

A simple explanation of the kinematic theory states that a rigid body in free space has six degrees of freedom, three of translation and three of rotation. Therefore, if three axes in space are chosen, as illustrated in figure 3, it is possible to move a body from one point to any other entirely by movements parallel to the three axes X, Y and Z. It is convenient to choose them to be mutually perpendicular. Any change in attitude can be achieved by suitable rotation about these axes. Therefore,

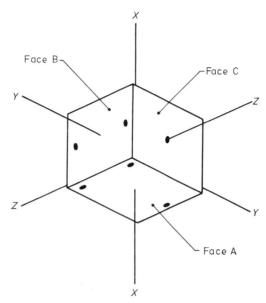

Figure 4 Position of six locators required to fully constrain a solid object.

any motion of a rigid body in free space may be resolved into three translations parallel to the co-ordinate axes, and three rotations about these axes.

From this one may derive the basic principle of kinematic design regarding the relative location of two component parts in a mechanism. This is that the number of points of contact and the number of degrees of freedom between two mating parts should always total six. They cannot be less, and any more means that some duplicate the function of others and are therefore redundant.

Figure 4 illustrates the six points of contact required to constrain fully a solid object, which for simplicity is shown as a cube.

The three points of contact on face A define a plane which destroys three of the six degrees of freedom, namely translation parallel to the axis X and rotation about the Y axis and the Z axis.

If there are two points of contact on face B two more degrees of freedom are removed, one of translation parallel to the Y axis and one of rotation about the X axis.

Lastly the one remaining degree of freedom, translation along the Z axis can be constrained by a single location point at the centre of face C.

The cube is thus constrained completely by the six locators shown.

It is possible therefore to classify a pair or coupling in a mechanism by the number of degrees of freedom.

Leaving couplings for the moment with no degrees of

freedom we have those with one degree of freedom which are by far the largest class of coupling. The two members of a slide, or a bearing in its journal with no end float, are examples.

Couplings with more than one degree of freedom are seldom required. The bearing in its journal with the end float restraint removed and a Hooke's joint are examples of couplings with two degrees of freedom.

A tripod resting on a plane surface is an example of a coupling in which there are three degrees of freedom and a knife edge resting on a plane surface has four.

Only one constraint is required for a coupling with five degrees of freedom and a sphere resting on a plane surface fulfils this requirement with the proviso that the ball is held in contact with the plane surface by gravity or some other restraining force.

A body in free space, such as a projectile in flight has six degrees of freedom and therefore cannot be classified as a coupling.

5 Couplings with no degrees of freedom
Couplings with no degrees of freedom constitute the fixtures in a machine.

One of the earliest forms of kinematic fixture was devised by Lord Kelvin and is known as the 'Kelvin clamp' or the 'hole, slot and plane' fixing. It is shown in figure 5 at (a). In this arrangement the upper element has three hemispherical feet one of which rests in a trihedral hollow H, another in a V groove S which is in line with the trihedral hollow and the third on the flat surface P. There are thus six bearing points, three in the trihedral hollow, two on the faces of the V groove and one on the flat surface, whilst gravity can be used to hold the two parts together. They are thus located relative to each other with no degrees of freedom and no unnecessary constraints. Once positioned a foot of the upper member will

remain in position in the trihedral hollow H and any dimensional changes in the elements due to thermal variations will simply cause the other two feet to slide along their respective V and plane locations, and if the friction at these points is not too great, any resulting strains will be small.

An alternative arrangement is shown in figure 5(b) whereby the hemispherical feet in the upper elements engage in three V grooves, set at 120° spacing in the lower element. Thus again there are six points of contact, two in each V.

This arrangement is commonly used in instruments where the vertical axis of the upper element is required to retain its position relative to the axis of the lower element irrespective of thermal changes. It also enables the upper element to be removed and replaced in its original position precisely.

One of the problems with the Kelvin clamp at figure 5(a) is the production of a trihedral hollow. A conical hollow is often used but this is not sound kinematically, unless the mating foot is machined away to provide only three points of contact.

There are two fairly simple ways of producing the trihedral hollow which have been used successfully in many instruments. In the first, as shown in figure 6, three inclined surfaces are produced on the end of a cylindrical rod or block by using an angled cutter, usually 45°, to make three successive cuts across the diameter of the rod or block and rotating it through 120° for each cut.

The second method shown in figure 7 is to press three balls into a hole or ring to give three contact points for a hemispherical foot. The size of the hole may be calculated as

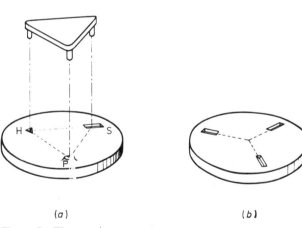

(a)   (b)

Figure 5 Three point support.

Figure 6 Method of forming a trihedral hollow.

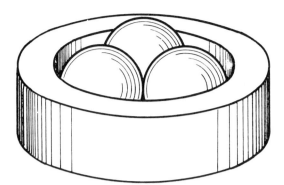

Figure 7 Balls used to provide three point support.

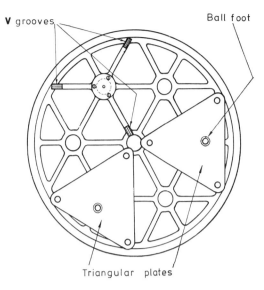

Figure 8 Kinematic support of large surface plate.

$D = d(1 + \sec 30°)$ where D is the diameter of the hole and d is the diameter of the balls.

An extension of the Kelvin clamp is sometimes used where large heavy weights, such as surface plates, have to be kinematically supported. Here, as in figure 8, the supports are arranged in multiples of three, each group of three reacting through a triangular link into a single reaction, which again may be in a group or groups of three. This system has been used successfully to support the mirrors on giant telescopes.

Where it is not possible to utilise gravity to maintain the parts in contact a spring or resiliant clamp may be used, but it is of the utmost importance that none of these springs or clamps should define a position for the body on which they act, but to apply a directional force only. In a coupling of a ball and a plane, the plane may be the fixed element and the ball be joined to the moving part; the reverse arrangement is possible and this often gives more freedom in the design. The point contact like the ball–plane and the ball–ball contacts are not the only possibilities; the contact between two crossed cylinders can also be useful.

6 Sliding members

If the upper member of two component parts is required to slide along a straight path over the lower member, i.e. with one degree of freedom, the V groove and trihedral hollow in the Kelvin arrangement can be replaced by a long V groove to accommodate two of the spherical feet and the third foot rests on a plane surface. The accuracy of the motion will depend on the uniformity of the V groove, its straightness and its parallelism to the plane of the surface which supports the third foot.

A simple and effective method of forming a guideway, without resorting to high precision machining is shown in figure 9. In this the guideway is formed by two hardened steel rods resting on a flat ground surface or supported in such a

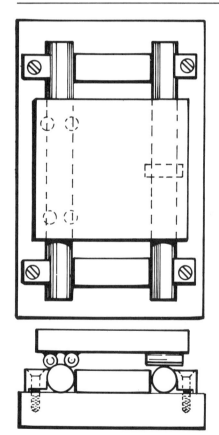

Figure 9 Kinematic slide using rods.

way that they are in the same plane. They must be straight but need not be of the same diameter and they should be suitably clamped down with a spacer between their ends. The upper member or carriage has on its underside four feet with spherical ends, balls may be used, forming in effect two V's locating on one of the bars. The other foot is formed by a cylinder fixed to the upper member so that it rests across the other guideway bar. There are thus five constraints with one remaining degree of freedom along the length of the guideway bars.

An important feature of this kinematic design is that if desired, as is often the case, the carriage may be removed from its ways and replaced precisely to follow its former path.

When hardened steel rods and balls, are used for this type of carriage it is a comparatively simple matter to correct for wear by rotating these parts in their mountings, and so restore the trackway to its former accuracy.

If the effort to drive such a carriage proves more than is desirable this can be minimised by replacing the sliding motion with one embodying steel balls rolling in three V grooves and on one flat surface as shown in figure 10.

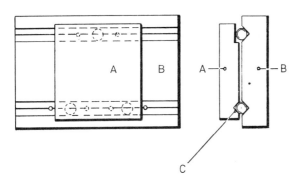

Figure 10 Kinematic slide using V's.

An alternative method of locating the cylindrical rods to form V grooves is to mount them in rectangular slots machined in both the base and the sliding carriage. Pairs of rods are located in each of the slots, the sides of which must be machined parallel to each other. For high precision work the rods should be ground and lapped with the carriage running on steel balls as above. The weight of the carriage serves to force the rods into contact with the sides of the rectangular grooves and hence maintain their alignment.

An important point to remember is that when the upper carriage has to travel a long distance along the guideways the use of balls may be unsuitable since they travel less than the moving carriage and this may produce instability of the carriage. In such cases the balls may be replaced by ball races

which have their periphery ground to a large radius, as in a self-aligning bearing, to give a point contact to accommodate any possible mal-alignment with the track.

Other important couplings having a single degree of freedom are the ball and V groove combination, and two threads (or wires) in a V configuration with the upper ends fixed at the same distance to each other and the lower ends joined to the moving part.

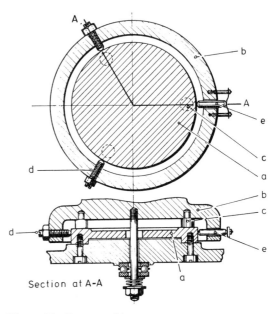

Section at A-A

Figure 11 Rotary table.

7 Kinematic principles applied to rotary members
In many cases accuracy of rotation is of great importance. In a plain bearing reducing the clearance to a minimum whilst still permitting smooth movement can nevertheless give rise to erratic errors in precision instruments.

To overcome this coupling can be constructed using slightly tapered cones with end restraint. The male and female elements are carefully machined to the same taper and fitted by scraping, which requires great skill in manufacture, or by lapping which cannot be done with high precision as a conical surface has a variable curvature along its length and rings tend to be produced along the lapped surface.

A simple method by which these objections may be overcome is shown in figure 11. A steel cylinder a is lapped on its upper face into a precision plane, the centre portion having been relieved for ease of lapping, and its cylindrical surface is ground and lapped to form as true a cylinder as is possible, its exact diameter being unimportant. The rotary top table b of the coupling is fitted with three pads c spaced at 120° which make contact with the base member at its lapped upper face. Two pads d spaced at 120° on the table top contact the cylindrical surface of the base member and a third spring loaded plug e fitted with an easy sliding fit in the top b ensures that contact is always maintained between the pads d and the cylindrical surface. Any clearance in the sliding fit of the spring plunger is not detrimental to the precise movement of the coupling.

As with linear slides the sliding constraints in a rotary coupling can be replaced with steel balls in V or ball races to reduce friction.

8 Screw couplings

Having kinematically designed a coupling for a precision instrument which permits relative motion between the two component parts it may be necessary to add a coupling between these parts which will accurately control this motion.

Of the many ways in which such a control may be accomplished the screw coupling is undoubtably the most common.

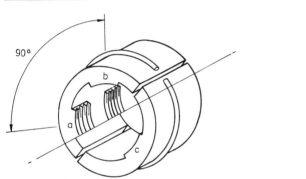

Figure 12 Split nut.

Errors arising from the various methods of constraining the axial movement of the screw are discussed later but of first importance is the accuracy of the actual screw thread.

The constraints required to locate a screw coupling precisely can be achieved by splitting the nut into two halves along its length (figure 12) and relieving one half to leave a pair of threaded areas a and b at each end, each pair being spaced at 90°. The other half of the nut has the thread cleared to leave two central areas c, so positioned that each of these areas is in line with and bisects the angle formed by two of the pads in the other half nut. When the two halves of the nut are assembled around a screwed spindle, and with a light spring to keep them together, the half nut containing the two central pads acts only as a force closure to keep the threaded spindle in contact with the four pads on the other half of the nut. If the forces on the coupling are extremely light it is possible to

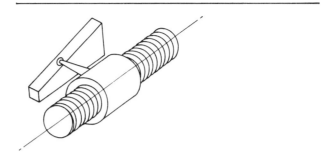

Figure 13 Nut with corrector bar.

dispense with the half nut with two pads c and use only the half nut with the four locating pads resting upon the threaded spindle to transmit the required motion.

Screw couplings constructed this way have practically no clearance therefore practically no backlash. The presence of backlash is usually considered undesirable but if it is present it is usually possible to introduce unidirectional thrust to ensure that the nut operates against one flank of the thread and, providing the setting of the screw is always made so as to oppose the thrust, the presence of a limited amount of backlash has no detrimental effect on the accuracy of the instrument.

In many instruments corrector bars are fitted to compensate for residual pitch errors in their measuring screws, which are usually progressive in nature, and in such cases it is unnecessary to have a highly accurate screw thread.

In the usual form of nut and screw assembly the nut is prevented from rotating with the screw by an arm attached to it and spring loaded or weighted against a bar (figure 13). This 'corrector bar' is adjusted so as to allow a slight rotation of the nut as it traverses the length of the screw, thus correcting for any general progressive errors in the screw. However care should be taken to avoid any irregularities, such as periodic errors, during the manufacture.

In this connection it may be interesting to mention a method of producing a nut which is contrary to kinematic principles. It is generally known as the principle of 'over constraint and elastic averaging', and was invented by Sir Thomas Merton.

This assumes that small errors exist in a screw and that these can be eliminated by elastic constraints. A split nut is produced as above, but in this case, the half nut with the four pads is produced from a slightly flexible material. In the early days a lining of pith was used but today new materials such as PTFE are more readily available. Due to the flexibility of this lining the traverse of the nut is dictated by the average pitch of the lead screw over the length of the nut, and the effect of periodic error of the lead screw on the accuracy of traverse of the nut is largely eliminated. A high order of accuracy can be obtained by producing a second, or even a third, screw from the corrected screw and this was the method originally adopted by NPL for producing diffraction gratings.

17

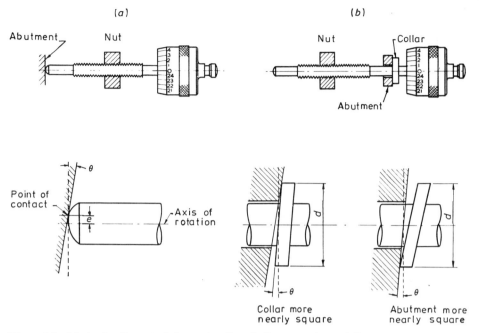

Figure 14 Methods of constraining screw threads from moving axially.

9 Errors in abutment faces

Following the design of kinematic screw couplings another very important factor is the method of employing coupling in precision instruments for sliding or rotating pairs. In general there are two cases to be considered, the first in which the screw is constrained from moving axially, and the second in which the nut is fixed.

In the first case the necessary constraints can be applied to the end of a screw as shown at (*a*) or against a collar integral with the screw, as at (*b*) in figure 14.

In the arrangement at (*a*), providing that the abutment face is accurately square to the axis of the screw and the abutment point at the end of the screw lies precisely on its axis there will be no periodic to and fro motion of the screw as it revolves. However if it is not square a periodic error, with an amplitude of $\pm e \tan \theta$, will be superimposed on the traverse of the nut as indicated in figure 14.

The situation is similar if a thrust collar is used as shown at (*b*). If the face of either the collar or the abutment is truly square there will be no axial motion of the screw when rotated. Should both be inclined however a periodic axial error will occur, the magnitude of which will depend on the diameter of the collar and its inclination of the face and will vary according to $\pm \frac{1}{2} d \tan \theta$.

In the case where the nut is fixed and the traverse of the screw is communicated directly to a carriage, or some other moving part, the arrangements shown at (*a*) and (*b*) are still

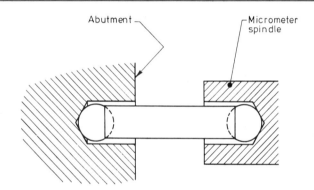

Figure 15 Ball ended strut.

applicable in that the abutment now forms part of the moving carriage and the nut is fixed to the stationary part.

An arrangement which overcomes the difficulty of period errors caused by malaligned abutment faces is shown in figure 15. In this the contact between the screw and the abutment face is through a ball ended strut which fits loosely in countersunk holes in both members. True alignment of the holes in the abutment and in the screw spindle is not essential as long as the strut is made fairly long.

10 Semikinematic design

So far only pure kinematic designs have been dealt with but

18

sometimes a compromise has to be made with a departure from the kinematically ideal for reasons to suit conditions of load, stability, robustness or ease of manufacture.

There is inevitably an increase in the demand on the accuracy of workmanship when a pure kinematic design cannot be used, but even semikinematic design usually shows a considerable saving on what may be termed traditional engineering methods.

Even with a pure kinematic design where theoretically one has a point contact the effect of an indentation of a ball location cannot be ignored.

In the case of linear translation, for example, the point contact on cylinders or balls rolling in Vs will not carry much load without serious loss of accuracy due to indentation or deformation of parts. Thus in instruments required for heavier duties the point or line contacts must at times give way to plane surfaces. Moreover in use these surfaces some wear will occur, flats will be worn on the bars forming the guideways and indentations will appear in the V grooves. Quite often the wear problems in V grooves can be minimised by lapping the V with a bar of the same diameter as the cylindrical bars or balls that will be used. By this method the accuracy of the initial alignment need not be as high; and the contact area can be greatly increased, with very little loss of accuracy.

This method of approach to the design of couplings is known as 'semikinematic' and may be defined as follows: if the area of each locator in a coupling is reduced to a theoretical point contact the design should become one of pure kinematics and not contain over constraints.

In the design of an instrument which is required to carry a fairly heavy load one should first visualise it designed on pure kinematic principles to determine the correct position for the locators, then it should be developed into a semikinematic design by expanding the theoretical point contacts into areas sufficiently large to carry the required load. In this way the possibility of producing over constraints is avoided.

11 Lobing of balls and cylinders

Balls and cylinders are used extensively for the guideways of precision measuring instruments and it is possible for these to contribute, in some degree, to the inherent errors in a measuring machine of they are not carefully selected to ensure freedom from lobing before assembly.

Figure 16 shows how it is possible for an apparently round roller, when measured between flat parallel anvils, to have a lobed form when rotated in a V block. This will be evident if the roller is rotated in a lathe against an indicator. Any lobing is not always regular in form and always appears with an odd number of faces and it arises more frequently when cylindrical work is ground on a centreless grinding machine.

Balls may also suffer from lobing, which is very difficult to eliminate other than by highly skilled lapping.

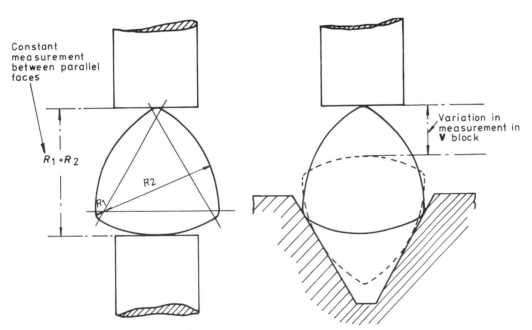

Figure 16 Lobing of balls and cylinders.

12 Deflections of members in instruments

Another factor in the mechanical design of precision measuring instruments which produces innaccuracies is the flexure of members within the instrument, and this applies particularly to the base, or bed, on which the measuring mechanism is mounted.

Careful attention must be paid to its stiffness in order to minimise distortion when under the influence of, not only the dead weight of the parts of the machine, but also of any variations in loading due to movement of slides or carriages

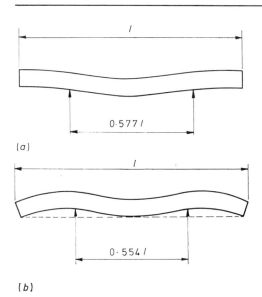

Figure 17 Supports for end bars and straight edges.

along the bed. The effect of mounting large and heavy work-pieces on a machine for measurement should also be borne in mind, as it is pointless in producing an instrument bed to a high degree of flatness if it will only distort when in use. This can be avoided by supporting heavy work-pieces independent of the table carrying the measuring equipment or mechanisms.

It is of the utmost importance that long work pieces are correctly supported to minimise possible errors that could arise from their deflection during measurement.

A former Astronomer Royal, Sir George Airy, supported the Imperial Standard Yard on a frame of eight rollers spaced apart according to $l/(n^2-1)^{1/2}$ where l is the length of the bar and n is the number of rollers.

It was subsequently found that two points of support spaced, centrally along the length of the bar, at a distance of $0 \cdot 577l$ were adequate to maintain the parallelism of the end faces. These points of support are known as the Airy Points, and are used for supporting end standard bars.

This however does not apply to a support for a straight edge since this requires a support giving a minimum of deflection of the straight edge. The correct separation of the supports to obtain this condition is $0 \cdot 554l$, which results in the sag at the ends being equal to the sag in the middle, as shown at (b) in figure 17.

These supports, when used in certain types of gauging and measuring equipment, can increase and maintain the accuracy of measurement and in many cases will also permit a lighter construction than would otherwise be practicable.

13 Conclusions

In conclusion, the advantages gained in adopting the principles and practices mentioned above when designing precision, mechanical parts of measuring instruments may be summarised as follows:

(i) The minimum number of contacts should be used to constrain each element in an instrument.

(ii) The forces at each contact should be known and controlled.

(iii) Elastic disturbance should be reduced to a minimum.

(iv) Each element need not be so accurately dimensioned.

(v) Parts may be made lighter, thus reducing wear through the use of less force.

(vi) Disturbing forces do not permanently affect the location of the parts since when removed the parts will revert to their initial positions.

A lack of appreciation of these principles in the design of precision mechanical parts of measuring instruments will surely lead to bad designs. A 'good design' can go a long way to annul bad workmanship but good workmanship can never save a bad design.

References

Braddick H J J 1956 *The Physics of Experimental Method* (London: Chapman and Hall)

Geary P J 1960 Torsional devices *BSIRA Report R249*

Pollard A F C 1929 *The Kinematic Design of Couplings in Instrument Mechanisms* (London: Hilger and Watts)

Whitehead T N 1954 Design and the use of instruments and accurate mechanisms *Notes on Applied Science No 15* (New York: Dover)

BS 1790: 1961 *Specification for Length Bars and their Accessories* (London: HMSO)

Chapter 3

Displacement transducers based on reactive sensors in transformer ratio bridge circuits

A L Hugill

The School of Earth Sciences, The Flinders University of South
Australia, Bedford Park, SA, 5042, Australia

Abstract. Displacement transducers which employ differential
reactive sensors in transformer ratio bridge circuits are
considered. Archetype models of inductive and capacitive
sensors are developed and the performance of bridge circuits
containing them is analysed. Aspects of performance which are
examined include linearity, discrimination, excitation forces,
power dissipation and the effects of stray impedances. Practical
design problems relating to long-term stability and
environmental influences are also discussed.

It is concluded that capacitive transducers are most suitable
in applications requiring high levels of accuracy, stability and
discrimination and low power dissipation and excitation forces.
In situations where these requirements are not so stringent both
inductive and capacitive transducers are suitable.

1. Introduction

Electrical displacement transducers are divided into four
functional elements by Garratt (1979). These are the mechanical
coupling, mechanical-to-electrical conversion, signal processing
and output elements. The mechanical coupling between the
transducer and the measurand and the conversion of the
resulting displacement are the functions most important in
determining performance. Two similar and widely used classes
of mechanical-to-electrical conversion elements are those in
which a displacement changes the inductance or capacitance of
a sensor. This change is then converted into an electrical signal.
Electrical energy for the signal is generated by the input
displacement in some sensors. These are termed self-generating
sensors or transducers. In others, called passive sensors, an
external electrical power supply is required. Passive inductive or
capacitive sensors can be connected into several different types
of circuits, for example oscillators or AC potentiometers.
However, in many situations the best configuration is to have a
differential or 'push–pull' type sensor forming the variable arms
of a bridge circuit. It is this configuration which is the subject of
the present discussion.

Aspects of the design and performance of reactive
displacement sensors in bridge circuits have been discussed by
many authors, for example Neubert (1975), Jones and Richards
(1973), Dratler (1977), Sydenham (1972) and Garratt (1979).
This paper is not intended to review this work but rather to use it
to help bring out in a systematic and comprehensive way the
similarities and differences between inductive and capacitive
systems. It is hoped that the reader will gain an understanding of
the way in which important design variables influence
transducer performance. Aspects of performance which are
dealt with include discrimination, which is related to electronic
noise, excitation forces due to the external power supply,
linearity, power dissipation and the effect of stray impedances.

2. Description of sensors

A differential reactive sensor consists of two capacitances or
inductances. An input displacement increases one and decreases
the other. When it forms the variable arms of a bridge circuit a
differential sensor has a major advantage. This is that both these

arms are close together and have nearly the same response to environmental changes, greatly increasing transducer stability. Other advantages are improved linearity and reduced excitation forces.

For some devices such as proximity detectors it is not possible to have a differential sensor. Here a bridge would consist of the ratio arms, a standard and the variable capacitance or inductance of the sensor. Although this paper deals with differential sensors most of the theory developed can be readily applied to single-element reactive sensors in bridge circuits.

2.1. Differential capacitive displacement sensors

Consider two overlapping parallel flat plates separated by a gap g. Let the area of overlap between them be A. If fringing fields are neglected the capacitance between the plates is $C = \varepsilon \varepsilon_0 A / g$ where ε is the relative permittivity of the substance in the gap and ε_0 is the permittivity of free space. Such a capacitor may be used as a displacement sensor if the displacement can be made to vary one or more of ε, A or g in a well defined manner. Usually g or A is variable to give a variable gap or variable area sensor. A differential sensor can be readily constructed using one of the three electrode configurations shown schematically in figures 1(a) and (b). The position of each of the two outer electrodes is fixed with respect to the other, and a third, common, electrode moves relative to these.

The capacitances C_1 and C_2 of the variable gap sensor in figure 1(a) are:

$$C_1 = \varepsilon \varepsilon_0 A / (x_0 - x) = C_0 x_0 / (x_0 - x) \qquad (1a)$$

and

$$C_2 = \varepsilon \varepsilon_0 A / (x_0 + x) = C_0 x_0 / (x_0 + x) \qquad (1b)$$

where x is the displacement of the common electrode from the central position. When $x = 0$ both electrode spacings are x_0 and the capacitances are C_0.

Similarly the capacitances for the variable area sensor shown in figure 1(b) are

$$C_1 = C_0 (x_0 + x) / x_0 \qquad (2a)$$

and

$$C_2 = C_0 (x_0 - x) / x_0. \qquad (2b)$$

Another variable area configuration is shown in figure 1(c). The earthed inner plate acts as a screen between the outer plates varying their effective area. In this case the common electrodes and the isolated outer electrodes are all fixed in position relative to each other. The expressions for C_1 and C_2 are given by equation (2).

The simple, or archetype, models described above do not account for the fringing field around the edges of electrodes. Therefore there will be some inaccuracies when they are applied to real sensors. Jones and Richards (1973) show that for

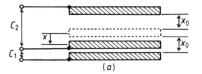

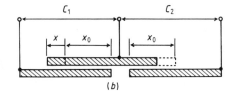

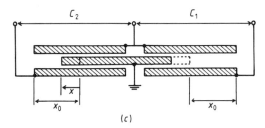

Figure 1. Capacitive sensors: (a), variable gap; (b), variable area; (c), variable area with grounded moving electrode.

electrodes with an area of about $300\ \mathrm{mm}^2$ and a gap of approximately 0.05 mm the fringing capacitance is between 3% and 5% of the total capacitance depending on the shape of the electrodes. This percentage will increase as the spacing between the electrodes becomes larger. With some added complexity the effect of fringing can be reduced by the use of a guard electrode such as described by Morgan and Brown (1969).

Sensors are not restricted to the parallel flat plate designs described above. Designs utilising spherical and cylindrical electrodes are reviewed by Sydenham (1972). The majority of differential variable gap sensors do, however, consist of parallel flat electrodes and their performance can be described approximately by equations (1). Variable area sensors of flat plate, co-axial cylinders (e.g. Wolfendale 1968) or more complex designs, such as described by Makow (1966) all perform, to a first approximation, according to equations (2). The capacitance C_0 of both types of sensors is typically between 1 pF and 100 pF.

2.2. Differential inductive displacement sensors

The principles by which a displacement can be made to vary the self-inductance of a coil have been summarised by Neubert (1975). The one most commonly used is the variation of reluctance. Another technique is the variation of the coupling between a coil and a conductor which is often a sleeve over the coil. There is another type of inductive sensor in which a displacement changes the mutual inductance between coils. In its most common form it is called the LVDT (linear variable differential transformer). As this sensor is not used in a bridge circuit it is beyond the scope of the present discussion.

Variable reluctance sensors can be divided into two groups. When the displacement changes a short gap in a high-permeability circuit the device is termed a variable gap sensor. Members of the other group have a plunger-type ferromagnetic armature penetrating a helical coil. These are termed VLP (variable leakage path) sensors because displacement changes the leakage path of the coil.

A variable gap sensor is shown schematically in figure 2(a). It is a differential configuration consisting of two cores and an armature, all made from high-permeability material. Each core has a winding of n turns and is in a fixed position relative to the other. The armature completes the magnetic circuits of the fixed cores and is able to move relative to them in the x direction. Let

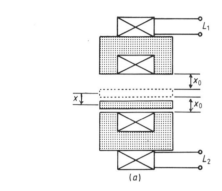

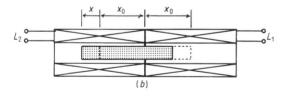

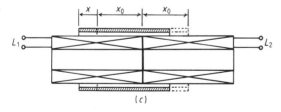

Figure 2. Inductive sensors: (a), variable gap; (b), VLP; (c), eddy current.

the gaps have an area A and a permeability μ_0. It is convenient to divide the flux in the magnetic circuit into three types depending on the path it follows. That which does not link all the turns on the coil is termed leakage flux. Of the flux which does completely link the coil, that which passes directly across the airgap will be called useful flux and the remainder is known as leakage flux. If the permeability of the cores and armature is high enough that the gaps form the dominant reluctance and fringing and leakage fluxes are neglected, then the inductances of the coils are

$$L_1 = \tfrac{1}{2}\mu_0 A n^2/(x_0 + x) = L_0 x_0/(x_0 + x) \qquad (3a)$$

and

$$L_2 = \tfrac{1}{2}\mu_0 A n^2/(x_0 - x) = L_0 x_0/(x_0 - x). \qquad (3b)$$

In these equations x is the displacement of the armature from the central position. When $x = 0$ the gap lengths and coil inductances are x_0 and L_0.

The ratio of fringing flux to useful flux depends on the dimensions of the sensor and decreases as the ratios of gap area to gap perimeter and gap length increase. Hugill (1981) describes a sensor made from the two halves of a 14 mm pot core with a gap x_0 of about 0.2 mm. In this case it is estimated that about 30% of the total inductance results from fringing flux and 7% from leakage flux. These inductances change more slowly with armature displacement than that due to useful flux.

The effect of fringing fields will be greater in inductive sensors than in capacitive sensors of approximately the same size. For a given total sensor area and gap, x_0, the ratio of gap area A, to x_0 will be at least twice as large for the capacitive sensor. This is because the inductive sensor, in practical form, must have a magnetic circuit with at least two gaps. Also there is no magnetic technique equivalent to the guarded electrode.

A simple analysis of a VLP sensor is more complex than that of variable gap sensors as it involves leakage flux distribution. Neubert (1975, p 182) gives an approximate analysis of a sensor the same as that shown schematically in figure 2(b). Using the results of this analysis the inductances of the coils, for a displacement x of the plunger, are

$$L_1 = L_0(x_0 - kx)/x_0 \qquad (4a)$$

and

$$L_2 = L_0(x_0 + kx)/x_0. \qquad (4b)$$

With the armature in the central position, $x = 0$ and the armature penetrates a distance x_0 into each coil. The constant k is less than unity and depends upon the sensor dimensions and effective armature permeability. It approaches unity for a well designed sensor.

These equations only hold true for a small linear range, typically less than $x_0/10$, and for x_0 about half the coil length. There is considerable mutual inductance, M_{12}, between the coils of the sensor in figure 2(b). For the small VLP sensor described by Hugill (1978) M_{12} is about half of L_0. The relative change in

M_{12} is less than 1% of the relative change in L_1 or L_2 over the linear range of the sensor. The coils of a VLP sensor may be decoupled by providing independent magnetic paths for each inductor.

The most effective way of utilising the coupling between a coil and a conductor in a displacement sensor is to partially surround the coil with a conductive sleeve. A differential sensor based on this method is shown schematically in figure 2(c). A high-conductivity sleeve partially surrounds two coils. This type of sensor has been analysed by Neubert (1975, p 187) and the inductances L_1 and L_2 as a function of sleeve displacement x are given by equations (4). The constant, k, will be less than unity and depends on geometric factors.

The archetype models described in this section do not take into account resistances and capacitances associated with the sensor. These depend on the design of individual sensors and it is difficult to make generalised comments on them. Their effects will be discussed in later sections of this paper.

The inductance, L_0, of typical sensor ranges from 1 mH to 100 mH.

3. Bridge circuits

3.1. Choice of bridge circuit

The value of C_0 for capacitive sensors is usually less than 100 pF and often less than 10 pF. If such a sensor is connected into a simple Wheatstone bridge, problems with stray earth impedances arise. The effect of these can be greatly reduced if a bridge with tightly coupled inductive ratio arms is used. This type of bridge is known as a Blumlein bridge or TRB (transformer ratio bridge). Theoretical and practical aspects of TRBs are discussed by many authors. This work is reviewed by Hague and Foord (1971).

The two basic types of TRBs are shown in figure 3, where Z_1 and Z_2 represent the sensor impedances. If the input voltage is connected across the windings as in figures 3(a) and (b) the transformer defines a voltage ratio and is termed a voltage ratio transformer or voltage transformer. If the detector is connected across the windings as in figure 3(c) and (d) a current ratio is defined. A transformer in this configuration is referred to as a current comparator or current transformer. Both the current and voltage transformers can have either two or three windings.

When used with a capacitive sensor the tapping point of the transformer is normally connected to earth. Stray capacitances which are often larger than the sensor capacitances exist between the bridge arms and earth. They originate from sources such as shielded cables and mechanical supports for sensor electrodes. An unbalanced stray capacitance as depicted by C_s in figure 3 will load one side of a voltage transformer. Because both windings of the transformer are linked by the same flux the ratio of the voltages across Z_1 and Z_2 remains constant at N_3/N_4. The magnitude of these voltages also remains constant if the input voltage generator has negligible impedance. With a current transformer the average flux density in the core is zero at

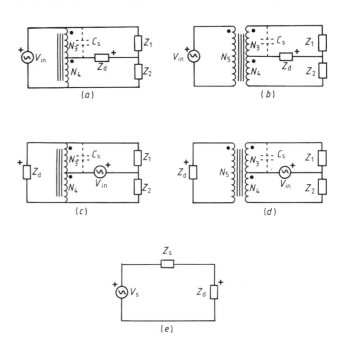

Figure 3. Transformer ratio bridges N_3, N_4 and N_5 are the windings on the transformers, Z_d the detector impedance and v_{in} the input voltage: (a), two-winding voltage transformer; (b), three-winding voltage transformer; (c), two-winding current transformer; (d), three-winding current transformer; (e), equivalent circuit of bridge.

bridge balance, and there is no voltage across C_s making it ineffective.

Apart from being able to detect small changes of capacitance in the presence of large stray capacitances TRBs have other desirable features which are listed below.

(i) The voltage or current ratio is very stable depending only on N_3/N_4 which is independent of temperature and time.

(ii) The ratio can be varied in accurate steps over a large range. When a capacitive sensor is used in a variable ratio bridge, range to resolution ratios of 10^6 or better have been reported by Morgan and Brown (1969), Stacey et al (1969) and Wolfendale (1968).

(iii) The ratio arms generate negligible thermal noise and can be designed to have very small power dissipation.

(iv) For many applications with displacement sensors the transformer can be of simple construction as it is stability of ratio rather than absolute ratio which is important. Jones and Richards (1973) use a three-winding transformer consisting of a loosely braided secondary hand wound on to a ferrite pot core with a copper screen between the primary and secondary windings.

24

(v) A three-winding transformer provides convenient isolation between the oscillator and detector.

Inductive sensors often have a low impedance and can be operated successfully in a Wheatstone bridge with resistive or uncoupled inductive ratio arms as described by Neubert (1975). If stray impedances due to causes such as long shielded cables or large sensor reactances were a problem a TRB would have to be used. To facilitate comparison between inductive and capacitive transducers this discussion will be restricted to sensors in TRBS.

3.2. Bridge sensitivity

For the purposes of comparison it is convenient to represent the bridge circuits of figures 3(a)–(d) in the form of an equivalent circuit. This is shown in figure 3(e) where the bridge arms and the oscillator are represented by a source voltage v_s in series with source impedance Z_s. When $N_3 = N_4 = N_r$ and the transformer is assumed to be ideal then the ratio arms will each have self-inductance L_r which will be the same as the mutual inductance M_r between the arms. In this situation the source voltage for both the voltage transformer bridges will be

$$v_s = v_0(Z_2 - Z_1)/(Z_2 + Z_1). \quad (5)$$

For the current transformer bridges

$$v_s = 2v_0(Z_2 - Z_1)/(Z_2 + Z_1 + Z_1 Z_2/Z_r) \quad (6a)$$

for the two-winding bridge and

$$v_s = [v_0(Z_2 - Z_1)N_5/N_r]/(Z_2 + Z_1 + Z_1 Z_2/Z_r) \quad (6b)$$

for the three-winding bridge. In these expressions the excitation voltage v_0 is the voltage across each side of the sensor when $Z_1 = Z_2$. So $v_0 = v_{in}$ in equation (6) and in equation (5) $v_0 = \frac{1}{2}v_{in}$ for the two-winding bridge and $v_{in}N_r/N_5$ for the three-winding bridge.

In equation (6) v_s depends upon the ratio of $Z_r = j\omega L_r$ to sensor impedance. If the sensor is capacitive then a current transformer bridge forms a resonant circuit. The performance of this type of circuit is examined by Neubert (1975, p 241). At operating frequencies below the resonant frequency the sensitivity (v_s/v_0x) of the bridge is proportional to $\omega^2 L_r C_0$. Operation in this region is undesirable as the excitation frequency and the magnitude of the ratio inductances must remain constant to ensure a stable output for a nonzero displacement. If the transducer is used only as a null detector then there could be some advantage in operating below resonance as the source impedance Z_s is approximately $4Z_r$ which is lower than the sensor impedance. Above resonance when $\omega^2 L_r C_0 \gg 1$ the sensitivity of the bridge becomes independent of ω and L_r. In most practical situations it is difficult to satisfy this condition and the bridge operates below resonance. In a typical example if $\omega = 10^5$, $L_r = 10$ mH and $C_0 = 10$ pF then $\omega^2 L_r C_0 = 10^{-3}$. If the ratio arms are shunted by a large constant capacitor as suggested by Neubert (1975,

p 242), stability is improved but sensitivity is reduced.

When the sensor is inductive the stability condition that Z_r is several times greater than Z_1 or Z_2 can be more easily satisfied. In this case the sensitivity of the current transformer bridge could be larger than the voltage transformer bridge. For a two-winding transformer the sensitivity will be twice as great. With the three-winding transformer any advantage depends upon N_5/N_r, which has the same limitations as following the voltage transformer bridge with a transformer of this ratio. As current transformer bridges suffer from problems with sensitivity and stability when used with capacitive sensors in practical situations and only offer a small advantage when used with inductive sensors the remainder of this paper will deal with sensors in voltage transformer bridges.

The source impedance of the voltage transformer bridges discussed above is

$$Z_s = Z_1 Z_2/(Z_1 + Z_2). \quad (7)$$

Equivalent source voltages and impedances for the different sensor types can be derived from equations (5) and (7) together with the expressions given in § 2. These are presented in table 1 where $Z_0 = j\omega L_0$ for inductive sensors and $1/j\omega C_0$ for capacitive sensors. They will be used in § 4 to determine transducer linearity and noise equivalent displacement.

3.3. Factors limiting bridge performance

Leakage inductance and series resistance in the ratio arms constitute the main limitation in the suppression of the effects of stray capacitances by TRBS. Leslie (1961) examines the results of these imperfections for several types of bridges. He concludes that there is little difference between two- and three-winding voltage transformers, providing the series inductances and resistances of the leads in the bridge are negligible. This is true for most capacitive sensors. For inductive sensors, however, they could be significant. If this were so and the ratio arm inductances, L_r, were larger than the sensor inductances L_0 then a two-winding transformer would be superior. The reason for this is that if the input voltage is connected across the bridge close to the sensor, then lead impedances appear in series with the larger inductances of the ratio arms, reducing their relative significance.

Jones and Richards (1973) estimate that for a typical capacitive sensor in a three-winding transformer bridge leakage inductance, including the leads, would be around 1 μH and stray capacitance about 1000 pF. If the bridge were operated at 16 kHz then an unbalanced change in leakage inductance or stray capacitance of 1% would be equivalent to a change in one of the sensor capacitances of 1 part in 10^7. As operating frequency is increased the bridge becomes more sensitive to stray and leakage reactances. To obtain equivalent stability with an inductive sensor and a two-winding transformer L_r would have to be around 10^5 times as large as the lead and leakage inductances. In the example above this means $L_r \geqslant 100$ mH, which would be difficult to attain. If a three-winding transformer were used the equivalent condition would be $L_0 \geqslant 100$ mH, which becomes harder to satisfy as sensor size decreases.

Therefore an inductive transducer would be at least as sensitive and in most cases more sensitive to changes in stray and leakage impedances associated with the bridge arms.

4. Associated electronics

When the moving element of a reactive displacement sensor is taken through the null position the output signal from the bridge detector changes phase by 180°. To determine the input displacement unambiguously the signal-processing stage of the transducer must be able to detect phase, relative to input voltage, and amplitude. A device which does this is termed a PSD (phase-sensitive detector) or lock-in amplifier. Other important features of PSDs are:

(i) Noise can be reduced as the amplifier acts as a band-pass filter which is 'locked' into the excitation frequency.

(ii) Quadrature signals are rejected. This is particularly important with inductive sensors which can produce a quadrature output due to resistive unbalance.

PSDs are described by Blair and Sydenham (1975) and Van Peppen (1978). They will not be discussed further in this paper as they are not normally the limiting factor in transducer performance.

The bridge detector is usually an operational amplifier in the inverting or non-inverting configuration as shown in figures 4(a) and (c). The output voltage of the detector, v_{out}, for the sensor models presented in § 2 are given in table 1. Assuming that the mechanical coupling element of the transducer has a linear characteristic two points can be noted.

(i) When a non-inverting amplifier is used any stray impedances, Z_g, shunting the detector will influence transducer sensitivity and some cases linearity. When $Z_g \gg Z_0$, as is mostly the case with inductive sensors, the influence is negligible and a stable linear output is obtained. With capacitive sensors care has to be taken to keep Z_0/Z_g constant and in the case of variable gap sensors as small as possible to increase linearity. To accomplish this the detector is mounted close to the sensor. Even so it would be difficult to reduce the ratio to less than one as there are stray earth capacitances associated with the mechanical supports of the moving electrode. The use of flexible shielded cable between the sensor and detector would reduce both the stability of any nonzero output and accuracy of the transducer calibration. If the strays are constant then nonlinearities can be reduced by electronic compensation in the form of a detector with negative input capacitance (Walker and Stroobant 1977).

(ii) These problems do not arise with inverting amplifiers. If Z_F is a capacitive reactance then the detector is termed a charge amplifier. When this type of detector is used with a capacitive sensor transducer sensitivity is independent of excitation frequency. Also flexible shielded cable could be used between sensor and detector without reducing stability or linearity. Similarly with an inductive sensor Z_F would have to be an inductive reactance to ensure that sensitivity was independent of excitation frequency.

The expressions for v_{out} in table 1 use the archetype sensor models described in § 2, and can only be used for a broad

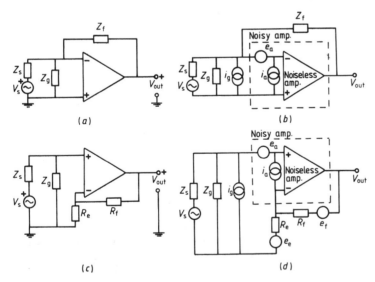

Figure 4. Detector circuits: (a), inverting amplifier; (b), noise equivalent circuit for (a); (c), non-inverting amplifier; (d), noise equivalent circuit for (c).

Table 1. Transducer performance for different combinations of sensor and detector.

	Capacitive sensors		Inductive sensors	
	Variable area	Variable gap	VLP, eddy current	Variable gap
v_s (source voltage)	$\dfrac{v_0 x}{x_0}$	$\dfrac{v_0 x}{x_0}$	$\dfrac{v_0 k x}{x_0}$	$\dfrac{v_0 x}{x_0}$
Z_s (source impedance)	$\tfrac{1}{2}Z_0$	$\tfrac{1}{2}Z_0[1-(x/x_0)^2]$	$\tfrac{1}{2}Z_0[1-(kx/x_0)^2]$	$\tfrac{1}{2}Z_0$
$-\tfrac{1}{2}v_{out}Z_0/Z_f$ (inverting amplifier)	$\dfrac{v_0 x}{x_0}$	$\dfrac{v_0 x}{x_0[1-(x/x_0)^2]}$	$\dfrac{v_0 k x}{x_0[1-(kx/x_0)^2]}$	$v_0 x/x_0$
$v_{out}R_e/(R_e+R_f)$ non-inverting amplifier	$\dfrac{v_0 x}{x_0[1+(Z_0/2Z_g)]}$	$\dfrac{v_0 x}{x_0\{1+(Z_0/2Z_g)[1-(x/x_0)^2]\}}$	$\dfrac{v_0 k x}{x_0\{1+(Z_0/2Z_g)[1-(kx/x_0)^2]\}}$	$\dfrac{v_0 x}{x_0[1+(Z_0/2Z_g)]}$
excitation forces: $Z_d \gg Z_0$	$-2G^2C_0 x$	0	0	$2G^2 x/\omega^2 L_0$
$Z_d \ll Z_0$	0	$\dfrac{2G^2C_0 x}{[1-(x/x_0)^2]^2}$	$\dfrac{-2G^2 x}{\omega^2 L_0[1-(kx/x_0)^2]^2}$	0

comparison of different systems. These models do, however, represent capacitive sensors more accurately than inductive sensors and the main limitations to stability and linearity of transducers based on the former will come from the effects of stray capacitance as described above. Linearity of inductive sensors, especially variable gap types will be limited by fringing and leakage fluxes rather than by the associated electronics. The amount of fringing flux depends on the design details of each sensor. The linearity of a small variable gap device is given by Hugill (1981).

Mutual inductance is greatest with VLP and eddy current sensors. These would normally be used with a non-inverting detector and $Z_g \gg Z_0$ to increase linearity. In this situation the expression for v_{out} in table 1 is multiplied by a factor $L_0/(L_0 \pm M_{12})$ to take into account the effect of mutual inductance. This factor is less than one for aiding fields and greater than one for opposing fields.

4.1. Electronic noise

The detector amplifier and any associated resistances are sources of electronic noise which is one of the factors limiting transducer resolution. Noise equivalent circuits of the two basic amplifier configurations discussed in the previous section are shown in figures 4(b) and (d). The 'noisy' amplifiers are represented by ideal noiseless amplifiers with noise voltage and current sources, e_a and i_a, at their inputs.

Any resistive component of Z_g produces Johnson noise represented by a noise current generator i_g. The feedback element Z_f of the inverting amplifier is assumed to be reactive and the feedback resistors of the non-inverting amplifier introduce noise voltages e_e and e_f.

All the above noise sources can be replaced by an equivalent input noise voltage generator, e_i, in the same position as the signal source v_s. Assuming noise sources are uncorrelated

$$e_i^2 = e_a^2|1 + Z_s/Z_g + Z_s/Z_F|^2 + (i_a^2 + i_g^2)|Z_s|^2 \qquad (8)$$

for the inverting amplifier and

$$e_i^2 = |1 + Z_s/Z_g|^2[e_a^2 + i_a^2|Z_g Z_s/(Z_g + Z_s) + R_e R_f/(R_e + R_f)|^2 + e_e^2|R_F/(R_e + R_F)|^2 + e_f^2|R_e/(R_e + R_f)|^2] + i_g^2|Z_s|^2 \qquad (9)$$

for the non-inverting amplifier. All noise voltages are in volts RMS/(Hz)$^{1/2}$ and noise currents in amps RMS/(Hz)$^{1/2}$.

4.2. Noise equivalent displacement

4.2.1. Capacitive sensors. Equating the expressions for v_s from table 1 to e_i gives the noise equivalent displacement,

$$x_n = e_i/G \qquad (10)$$

where $G = V_0/x_0$ is the responsivity of capacitive transducers and V_0 is the RMS amplitude of v_0. The minimum value of transducer discrimination will be x_n. It will be larger than this if there is any stiction or lost motion in the mechanical coupling element. For a given transducer the value of e_i depends upon the noise bandwidth of the filter in the PSD. Therefore, x_n, and potential discrimination are directly related to transducer bandwidth. A wider bandwidth leads to a larger value of x_n. The upper limit on bandwidth is about $\frac{1}{3}$ of the excitation frequency.

Letting $Z_s = 1/2j\omega C_0$ in equation (8), the equivalent input noise for a charge amplifier with feedback capacitor C_F is

$$e_i^2 = e_a^2[(C_f + C_g + 2C_0)^2/4C_0^2 + 1/4\omega^2 C_0^2 R_g^2]$$
$$+ (i_a^2 + i_g^2)/4\omega^2 C_0^2. \quad (11)$$

Here the earth impedance Z_g consists of a capacitive component C_g coming largely from leads and mechanical supports and a resistive component R_g coming from insulator losses and the resistance necessary to drain the amplifier input bias current.

If the non-inverting amplifier has a moderate gain, greater than 10 say, and the parallel combination of Z_g and Z_s is much larger than R_e, an approximate expression for e_i can be obtained from equation (9). This is

$$e_i^2 = (e_a^2 + e_e^2)[(C_g + 2C_0)^2/4C_0^2 + 1/4\omega^2 C_0^2 R_g^2]$$
$$+ (i_a^2 + i_g^2)/4\omega^2 C_0^2. \quad (12)$$

The feedback resistor R_e can make a significant noise contribution if care is not taken in design. For example if $R_e = 9$ kΩ then $e_e = 12$ nV/(Hz)$^{1/2}$ which is approximately equal to e_a for a low-noise IC op-amp.

For both amplifiers the term involving noise currents is reduced by keeping excitation frequency ω and R_g as high as possible, and using a FET input amplifier with a low noise current. The problem of suppressing noise currents becomes more difficult as C_0 decreases. Due to mechanical design constraints it is normally difficult to make C_g less than C_0 even if the amplifier is mounted close to the sensor. The noise voltage term involving R_g is normally much less than 1. Therefore the lowest equivalent input noise from a well designed practical system would be around $1.5e_a$ for the non-inverting amplifier and around the same value for the inverting amplifier if $C_F < C_0$.

For a typical variable gap sensor with $x_0 = 0.5$ mm and $V_0 = 10$ V and an IC op-amp with $e_a = 10$ nV/(Hz)$^{1/2}$, $x_n \approx 10^{-12}$ m/(Hz)$^{1/2}$. The responsivity of the transducer can be increased by decreasing x_0 or increasing V_0 until either the dielectric in the gap breaks down or the acceptable excitation force between the plates is exceeded. If the dielectric were air and $Z_g = \infty$, Dratler (1977) shows that x_n could be reduced to 5×10^{-15} m/(Hz)$^{1/2}$ before electrical breakdown occurs. Jones and Richards (1973) describe a system where e_i is reduced by connecting an inductance across the detector to form a resonant circuit. Such a system increases the complexity of the transducer and is susceptible to interference and detuning.

4.2.2. Inductive sensors. The noise equivalent displacement for inductive transducers is given by equation (10) where $G = V_0/x_0$ for variable gap sensors and kV_0/x_0 for VLP and eddy current sensors.

In practical transducers Z_s is relatively low for inductive sensors and the effect of Z_g and i_a can be neglected. If the voltage gains of both the inverting and non-inverting amplifiers are about 10 or greater, and $e_e^2 \ll e_a^2$ then an approximate expression for equivalent input noise in both cases is $e_i = e_a$.

This value could be obtained with no special precautions and would not normally be increased by the use of shielded cable between sensor and detector. With a capacitive transducer great care would be required to reduce e_i to around $1.5 e_a$. Despite this the ultimate value of x_n will not be as small as for capacitive transducers as responsivity will be limited by power dissipation and excitation forces.

For capacitive sensors with air gaps power dissipation is negligible, but with inductive sensors there has to be some resistance in the windings, and there will be losses in the high-permeability core or eddy-current sleeve. A general expression relating power dissipation to x_n would be complex as it depends upon the geometry, material properties and winding details of the sensor. The power dissipation of a typical sensor will be given in § 6. Excitation forces will be discussed in the next section.

5. Excitation forces

5.1. Capacitive sensors

The excitation voltage across the sensor electrodes produces electrostatic forces. In some applications, such as in miniature seismometers (Usher *et al* 1977), these forces can be a major factor limiting responsivity and hence discrimination. The force between two electrodes depends upon the rate at which the stored mechanical energy of the system changes with displacement. Consider a capacitor C with a voltage v across it. If one of the electrodes is able to move in an arbitrary x direction, then C will be a function of x. The magnitude of the force on the moving electrode in this direction is given by (e.g. Tagg 1974)

$$F = \tfrac{1}{2}v^2 \, dC/dx. \quad (13)$$

If v is a sinusoidal voltage then the instantaneous force consists of the sum of two components. One is sinusoidal with a frequency twice that of v, and the other is a steady force. In most displacement sensors the excitation frequency will be sufficiently high that the moving electrode will not respond to the oscillating component. The remaining steady component is given by equation (13) with the sinusoidal voltage v replaced by its RMS value.

With the differential sensors described in § 2 there are two forces in opposite directions acting on the moving element. When $x = 0$ the sum of these forces will be zero. For off-centre positions the net force will depend upon electrode position, the circuit in which the sensor is connected and its geometry. For sensors in two- or three-winding voltage transformer bridges two cases will be considered.

(i) $Z_d \ll Z_0$. This corresponds to the charge amplifier discussed in the previous section. Here the voltage across each half of the sensor is constant and the magnitude of the force on the moving electrode is

$$F = \tfrac{1}{2}V_0^2(dC_1/dx + dC_2/dx). \quad (14)$$

This force is in the x direction as defined in figure 1 and V_0 is the RMS value of the sinusoidal excitation voltage v_0.

(ii) $Z_d \gg Z_0$. This corresponds to the non-inverting amplifier with $Z_g \gg Z_0$. In this case

$$F = 2V_0^2(C_2^2\, dC_1/dx + C_1^2\, dC_2/dx)/(C_1 + C_2)^2. \quad (15)$$

Equations (14) and (15) are evaluated for both variable gap and variable area sensors and the results are presented in table 1.

5.2. Inductive sensors

The magnetic force on the armature of an inductive sensor depends upon the current through the sensor and the rate of change of inductance with displacement. The magnitude of this force due to a current i_1 in the coil of the inductance L_1 of any of the sensors illustrated in figure 2 is (e.g. Rotors 1941) $F = \tfrac{1}{2}i_1^2(dL_1/dx)$. This force is directed in the x direction. The current in inductance L_2 produces a similar force in the opposite direction. The net force on the armature is the sum of the two. For sensors in two- and three-winding voltage transformer bridges two cases are considered.

(i) $Z_d \ll Z_0$. This condition is satisfied when an inverting amplifier is used as the detector. The voltage across each coil will remain constant, independent of armature position, and the force on the armature is

$$F = \tfrac{1}{2}V_0^2\left(\frac{dL_1/dx}{\omega^2 L_1^2} + \frac{dL_2/dx}{\omega^2 L_2^2}\right) \quad (16)$$

(ii) $Z_d \gg Z_0$. This is satisfied when a non-inverting amplifier is used as the detector. In this case the current through both coils is equal and the force on the armature will be

$$F = 2V_0^2(dL_1/dx + dL_2/dx)/\omega^2(L_1 + L_2)^2 \quad (17)$$

where ω is the frequency of the excitation voltage.

Equations (16) and (17) are evaluated for all three types of sensors and the results presented in table 1.

5.3. Discussion

According to table 1 there will be zero force on the moving element of each of the sensor types described in § 2 if the appropriate detector impedance is used. The assumptions made in obtaining these results will only be valid in practice for the variable area, VLP and eddy-current sensors within their linear ranges. With the variable gap capacitive sensor it was assumed that $Z_g \gg Z_0$ which is difficult to achieve as discussed previously. As Z_g decreases the force on the electrode approaches that for the charge amplifier. For the variable gap inductive sensor the archetype model does not take fringing and leakage fluxes into consideration. If these are considered the force on the armature of the transducer with the inverting amplifier tends towards that with the non-inverting amplifier.

A realistic 'order of magnitude' comparison of the excitation forces in transducers with variable gap sensors can be made by using the relevant nonzero force expressions in table 1. For transducers with the same responsivity and input displacement the ratio of the force on a capacitive sensor to that on an inductive sensor is

$$K = \omega^2 L_0 C_0. \quad (18)$$

This expression is only generally valid for $x^2 \ll x_0^2$. For sensors with $L_0 = 5$ mH, $\omega = 5 \times 10^4$ and $C_0 = 10$ pF this ratio is about 10^{-4}. The force on the inductive sensor can be reduced by increasing ω, L_0 or both. The extent to which these modifications can be taken is limited in practice. If they could be increased until $K = 1$, then the impedance of the inductive sensor would equal that of the capacitive sensor operated at the same frequency. Any advantage resulting from the low impedance of the inductive sensor would be lost. Therefore in most situations the excitation force on variable gap capacitive sensors will be considerably less than those on inductive sensors of the same type.

6. Practical considerations

6.1. Capacitive sensors

If the dielectric between the active electrodes of a capacitive sensor is air or some other gas and the electrodes are good conductors then the capacitances C_1 and C_2 will have negligible losses. To accurately represent real sensors the archetype models in § 2 only have to be altered to take into account fringing fields.

Practical construction details are discussed by several authors, e.g. Jones and Richards (1973), Dratler (1977), Morgan and Brown (1969), Corner and Hunt (1954), Brown and Bulleid (1978) and Khan et al (1980). Sensor geometry is usually simple consisting of metal electrodes supported by solid dielectrics and factors controlling mechanical stability and the effects of temperature, humidity and pressure can be readily assessed. The impedance to ground from the outer electrodes has negligible effect on the performance of a well designed bridge as discussed in § 3. The supports for these electrodes will, however, be one of the major factors influencing the mechanical stability and temperature dependence of the transducer. Jones and Richards (1973) use thin (0.1 mm) sheets of optical quality mica leading to relatively large stray capacitances but favourable mechanical and thermal performance. Impedance to ground from the common plate has a greater influence on the

29

electrical performance of the transducer. This impedance shunts the detector contributing to Z_g, which limits transducer discrimination and in some cases linearity. Usher (1977) reduces capacitance shunting the detector to about 10 pF by using relatively thick quartz spacers as mechanical supports.

Environmental changes influence the sensor in two ways. They alter the electrical properties of the dielectrics and the mechanical dimensions of the electrodes and their supports. Dimensional changes, mostly due to changes in temperature, can, because of the simple sensor geometry, be readily assessed from the properties of the materials involved. The dielectric properties of a gas depend upon temperature, pressure and humidity. The effect of changes in these on a capacitor with an air gap is discussed by Bell (1960).

With differential sensors environmental influences can be reduced by symmetric design. A controlled environment would be necessary for transducer stability to approach the noise equivalent displacement x_n.

6.2. Inductive sensors

The archetype models used in this paper only consider coil inductances. In real sensors there are capacitances and losses associated with the windings and losses in the magnetic materials and conductors. It is difficult to generalise on these subjects as they depend on the design details of individual sensors and the frequency and amplitude of the excitation voltage. Detailed analyses of losses and self-capacitances in transformers and inductors are given by Snelling (1969) and Welsby (1960). Neubert (1975) applies some of these results to sensors.

If the losses are lumped together as a series resistance then the impedances in equation (5) becomes $Z_1 = R_1 + j\omega L_1$ and $Z_2 = R_2 + j\omega L_2$. The equivalent source voltage is

$$v_s = v_0 \{ (R_2 - R_1)/Q^2(R_2 + R_2) + (L_2 - L_1)/(L_2 + L_1)$$

$$+ j[(L_2 - L_1)/(L_2 + L_1)$$

$$- (R_2 - R_1)/(R_2 + R_1)]/Q \}/(1 + 1/Q^2). \quad (19)$$

The real part of this expression is the in-phase signal and the imaginary or quadrature component is rejected by the PSD. Resistive imbalance and changes in loss resistances will affect both the signal and quadrature terms. Although the quadrature term is eliminated, it will limit the amount of AC amplification which is possible before the signal is rectified. Thermal effects due to power dissipation could also restrict the amplitude of the excitation voltage and hence responsivity of the transducer. For these reasons it is desirable to make $Q = \omega(L_1 + L_2)/(R_1 + R_2)$ as high as possible.

Capacitances associated with each winding can be lumped together. The resulting self-capacitance appears in parallel with the inductance of the winding. It lowers the Q of the sensor and also the stability of the transducer if resonance is approached.

To illustrate the effects of losses and self-capacitances in a practical situation the sensor described by Hugill (1981) will be discussed. This is a variable gap type made from 14 mm ferrite pot cores with $x_0 = 0.225$ mm. The winding on each half of the sensor consists of 200 turns of 40 AWG enamelled copper wire having an inductance of about 4 mH, resistance of around 20 Ω and a self-capacitance of less than 10 pF. When the excitation voltage, v_0, is 1 V RMS at 20 kHz the dominant losses come from the DC winding resistance. Under these conditions the sensor Q is 25 and each coil dissipates about 0.1 mW. The resulting temperature rise will depend upon the thermal properties of the environment. A value of 0.01°C is obtained by using the manufacturers' specifications for the pot core surrounded by air.

The noise equivalent displacement can be obtained from § 4.3 and is about 2×10^{-12} m/(Hz)$^{1/2}$ when the detector has a noise voltage of 10 nV/(Hz)$^{1/2}$. To maintain dimensional stability equal to this the temperature of a 1 cm length of ferrite would have to be kept constant to within 2×10^{-5}°C.

The effect of resistive imbalance will depend upon the range the sensor is being used over. If this is ± 1 μm then according to equation (19) R_1 and R_2 would have to be balanced to within 0.2 Ω for the quadrature component to be reduced to about 10% of the maximum signal. This would probably require a resistive balance adjustment and the problem would become worse as the range is reduced.

As the resonant frequency of the sensor is about 1 MHz self-capacitance has negligible influence at 20 kHz.

The main environmental influences on inductive sensors are temperature and magnetic field. Variations in temperature change the dimensions and the magnetic and electric properties of the materials in the sensor. Fedotov (1974) evaluates the effects of these changes for variable gap type sensors. Changes in the ambient magnetic field can alter the relative permeability and dimensions of magnetic materials and also the force on the armature. Although the symmetry of a differential sensor reduces these environmental influences it would be necessary to control temperature and magnetic field for most high-accuracy applications. Humidity does not have a first-order effect on sensor performance.

Long-term stability of VLP and eddy-current sensors depends directly on the dimensional stability of the windings. This would be difficult to evaluate as it involves the properties of the wire, the insulation and the coil former, or adhesive in the case of formerless windings. These types of sensor would have a lower long-term stability than a well designed capacitive sensor.

The winding of a variable gap inductive sensor does not directly influence stability, although there are some second-order effects due to leakage. Here it is the dimensional stability of the gap which is important. This depends upon the stability of both the components in the magnetic circuit and their mechanical supports. For most practical sensors the magnetic materials available to the designer are restricted to laminated high-permeability metals or ferrites. Further design difficulties arise because of the relatively complex geometry required to house the coils and to complete the magnetic circuit.

The designer of capacitive sensors has a wider choice of materials and fewer restrictions on design because of the relatively simple geometry. For these reasons it would be very difficult to build a variable gap inductive sensor with long-term stability comparable to a capacitive sensor. To the author's knowledge there has been no experimental comparison of the stability of practical capacitive and inductive sensors.

7. Conclusions

It is difficult to make a simple generalised comparison between inductive and capacitive transducers because many aspects of performance depend upon the practical design details of individual systems. In particular sensor design involves the modelling of electric and magnetic field distributions. If this is to be done accurately it is a complex process involving a detailed knowledge of dimensions and material properties. To make a broad comparison of different sensor types the models used must involve simplifications. With these simplifications, some of the details on which the performance of real transducers depend are lost. The problem is made worse by the lack of experimental data comparing the two types of transducer. Several differences are however apparent for sensors of comparable size under typical operating conditions.

(i) The reactance of capacitive sensors is large while that of inductive sensors is small, leading to different design problems. With capacitive transducers any capacitance of shunting the detector decreases the signal-to-noise ratio. When a high-impedance detector is used it can also reduce transducer stability with a variable area sensor, and both stability and linearity with a variable gap sensor. Reduction of these effects requires careful design of the sensor and minimisation of the distance between it and the detector.

The low reactance of inductive sensors makes the bridge circuit sensitive to changes in stray series inductances associated with leads and the ratio transformer. In applications requiring high stability the ratio arms would have to be mounted as close as possible to the sensor to reduce these effects.

(ii) The linearity of capacitive and inductive transducers with variable gap sensors in fixed ratio bridge circuits are each limited by different processes. For capacitive sensors, particularly those with a guarded electrode, the limitation comes from stray capacitances as mentioned in (i) above. With inductive systems the limitation arises from leakage and fringing fluxes.

(iii) When variable gap sensors are used in transducers with the same responsivity the excitation force in the capacitive sensor will be less than in the inductive sensor.

(iv) Power dissipation in capacitive sensors with gas dielectrics is negligible. With inductive sensors it is one of the major factors limiting responsibility.

(v) There is more scope to design capacitive sensors which are stable with temperature and time. This is because they can be constructed from a wider range of materials and their geometry is simpler than inductive sensors. Also in some situations the stability of inductive sensors depends upon winding wire and its insulation.

The above points account for why capacitive transducers are used in the majority of applications where high levels of accuracy, stability and discrimination, and low power dissipation and excitation forces are required. Both inductive and capacitive transducers are suitable for applications where these requirements are not so stringent. Inductive transducers have an advantage when it is not convenient to locate electronics close to the sensor.

References

Bell D A 1960 Temperature coefficient of capacitance and of inductance
Electron. Techn. **37** 342–5

Blair D P and Sydenham P H 1975 Phase sensitive detection as a means to recover signals buried in noise
J. Phys. E: Sci. Instrum. **8** 621–7

Brown M A and Bulleid C E 1978 The effect of tilt and surface damage on practical capacitance displacement transducers
J. Phys. E: Sci. Instrum. **11** 429–32

Corner W D and Hunt G H 1954 A direct reading instrument for the measurement of small displacements
J. Sci. Instrum. **31** 445–7

Dratler J 1977 Inexpensive linear displacement transducer using a low power lock-in amplifier
Rev. Sci. Instrum. **48** 327–35

Fedotov A V 1974 Evaluating the temperature error of inductive measuring transducers
Measurement Techniques **17** 95–9

Garratt J D 1979 Survey of displacement transducers below 50 mm
J. Phys. E: Sci. Instrum. **12** 563–73

Hague B and Foord T R 1971 *Alternating Current Bridge Methods* 6th edn (London: Pitman)

Hugill A L 1978 Synthesis of inductive displacement-measuring system using computer-aided design
Proc. IEE **125** 417–21

Hugill A L 1981 Probable flux path modelling of an inductive displacement sensor
J. Phys. E: Sci. Instrum. **14** 860–4

Jones R V and Richards J C S 1973 The design and some applications of sensitive capacitance micrometers
J. Phys. E: Sci. Instrum. **6** 589–600

Khan A R, Brown I J and Brown M A 1980 The behaviour of capacitance displacement transducers using epoxy resin as an electrode-guard ring spacer
J. Phys. E: Sci. Instrum. **13** 1280–1

Leslie W H P 1961 Choosing transformer ratio-arm bridges
Proc. IEE **108** 539–45

Makow O M 1966 Precise measurement of length using capacitance ratio
Metrologia **2** 125–6

Morgan V T and Brown D E 1969 A differential-capacitance transducer for measuring small displacements
J. Phys. E: Sci. Instrum. **2** 793–5

Neubert H K P 1975 *Instrument Transducers* 2nd edn (London: Oxford University Press)

Rotors H C 1941 *Electromagnetic Devices* (New York: Wiley) Chap. 8

Snelling E C 1969 *Soft Ferrites* (London: Iliffe)

Stacey F D, Rynn J M W, Little E C and Croskell C 1969 Displacement and tilt transducers of 140 dB range
J. Phys. E: Sci. Instrum. **2** 945–9

Sydenham P H 1972 Microdisplacement transducers
J. Phys. E: Sci. Instrum. **5** 721–33

Tagg C F 1974 *Electrical Indicating Instruments* (New York: Crane Russak) pp 180–1

Usher M J, Buckner I W and Burch R F 1977 A miniature wideband horizontal-component feedback seismometer
J. Phys. E: Sci. Instrum. **10** 1253–60

Van Peppen J C L 1978 Coherent detection and its use in lock-in amplifiers
In *Modern Electronic Measuring Systems* ed. P P L Regtien (Delft: Delft University Press)

Walker I J and Stroobant A D 1972 Precision transducer systems utilizing differential capacitance measurements
JAEU (Japan) **5** 32–7

Welsby V G 1960 *The Theory and Design of Inductance Coils* 2nd edn (London: McDonald)

Wolfendale P C F 1968 Capacitive displacement transducers with high accuracy and resolution
J. Phys. E: Sci. Instrum. **1** 817–8

Chapter 4

Silicon micro-transducers

S Middelhoek and D J W Noorlag
Department of Electrical Engineering, Delft University
of Technology, Delft, The Netherlands

Abstract By means of silicon planar technology it is
possible to make not only integrated circuits but also
integrated sensors and actuators. This paper gives a
comprehensive review of research work in the sensor area.
It also includes some general considerations with respect
to measurement systems and signal conversion as well as
some conclusions and comments on a future outlook.

1 Introduction

Electronic measurement and control systems or, for that
matter, information processing systems in general consist, as
shown in figure 1, of an input transducer, a signal modifier
and an output transducer (Middelhoek and Noorlag 1980).
In the input transducer often called sensor, a measurand such
as a temperature, pressure or chemical concentration is converted into an electronic signal. In the modifier, called signal
processor, the electronic signal is in some way modified that

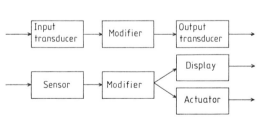

Figure 1 Functional block diagram of measurement and
control systems.

is, amplified, filtered or converted from an analogue signal
into a digital one. Even the central-processing unit of a digital
computer can be considered to be a modifier. In the output
transducer the electronic signal is converted into a signal
which can be perceived by one of our senses. When the
electronic signal is converted into an optical signal, such an
output transducer is called a display.

In control systems the electronic signal is also used to cause
some useful action such as igniting a gas flame, closing a
valve or printing a character. In that case the output transducer is called an actuator.

Until recently much of the research and development work
in industry was focused on the electronic signal-processing
part, so that today, thanks to the silicon planar technology, we
have at our disposal a huge number of very sophisticated LSI
components such as microprocessors, memories, A/D converters, operational amplifiers, etc. Since the price/performance
ratio of these components has drastically decreased during
the last decades, one is gradually becoming aware of the fact,
that the price/performing ratio of the input and output transducers is seriously lagging behind that of integrated circuits.
This might, in fact, determine the future feasibility of low-cost
electronic measurement and control systems and innovative
consumer products. To narrow this gap more and more
research and development is being devoted to the field of
transducers. One of the results of this increased activity is a
new class of transducers, called Silicon Micro-Transducers
(SMT's), which are more or less fabricated by the same silicon
planar technology used for the production of integrated
circuits (Middelhoek *et al* 1980).

The aim of this paper is to present a comprehensive review
of research work being carried out in this area. The main
section describing the most important devices is preceded by a
section giving some considerations with respect to the signal
conversion in transducers and the different signal domains
which are of interest. In the last section future trends will be
indicated.

33

2 Signal conversion in measurement systems

In an input transducer or sensor a physical quantity or, more generally, a nonelectrical signal is converted into an electrical signal. For instance, in a photovoltaic cell (solar cell) light intensity is converted into a voltage and/or an electrical current. If the light is modulated with some information, the electrical signal at the output of the input transducer will carry the same information. In fact energy is used as a carrier of information.

In the above photovoltaic cell signal conversion goes hand in hand with energy conversion. In order to obtain insight into the different signal conversions in a transducer, it is useful to consider the different forms in which energy manifests itself and to study the physical effects by means of which energy conversion can take place.

From a physical point of view, we can distinguish the following forms of energy (Van Dijck 1964):

(i) *Radiant energy* is related to radiowaves, microwaves, infrared, visible light, ultraviolet, x-rays and γ-rays;

(ii) *Gravitational energy* concerns the gravitational attraction between a mass and the earth.

(iii) *Mechanical energy* pertains to mechanical forces, displacements, flows, etc.;

(iv) *Thermal energy* is related to the kinetic energy of atoms and molecules;

(v) *Electrical and magnetic energy* deals with electric and magnetic fields, currents, etc.;

(vi) *Molecular energy* stands for the bond energy, with which atoms are held together in a molecule;

(vii) *Atomic energy* is the binding energy which is related to the forces between nucleus and electrons;

(viii) *Nuclear energy* is the binding energy which keeps the nuclei together;

(ix) *Mass energy* as proposed and described by Einstein.

In the search for alternative energy sources today all methods that allow efficient conversion from one of the energy forms into electrical energy are being studied. For measurement purposes it is convenient to group the energy forms into six main energy domains, which in turn lead to six main signal domains (Lion 1969): radiant signals such as light intensity or wavelength, mechanical signals such as pressure or level, thermal signals such as temperature or heat flow, electrical signals such as voltage or dielectric constant, magnetic signals such as magnetic-field strength or permeability and chemical signals such as composition or pH.

In an electronic measurement system one of the five non-electrical signals is converted in the input transducer into an electrical signal. The electrical signal is then processed in the modifier and the electrical signal is converted back into one of the five nonelectrical signal domains in the output transducer. Figure 2 shows a general diagram representing all possible electronic measurement systems.

Considering the different transducers and their effects it appears that two kinds of transducers exist. Some input

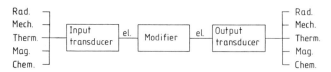

Figure 2 General diagram of an electronic measurement system.

transducers generate an electrical output without an auxiliary source of energy, e.g. a solar cell or a thermocouple. Such transducers are called *self-generating transducers*. In these the energy at the output is always smaller than the energy at the input of the transducer. Other transducers can only convert one of the five input signals into electrical energy when an auxiliary source of energy is available, e.g. a photoconductor, a strain gauge, a magnetoresistor, etc. Such transducers are called *modulating transducers*. Since an auxiliary source of energy is used, the energy level at the output can be much larger than that at the input of the transducer. Looking at output transducers that convert an electrical signal into a radiant signal it appears that a light-emitting diode (LED) is a self-generating transducer, whereas a liquid-crystal display is a modulating transducer because the radiant energy is modulated by a very small electrical signal. In figure 3 the graphic symbols are shown for both self-generating and modulating transducers, and a diagram is given to illustrate the use of the symbols to represent an electronic thermometer with a temperature-sensitive transistor as input transistor and a liquid-crystal display as output (Middelhoek and Noorlag 1981).

As said before the advancement with respect to the electronic modifiers has been enormous in recent decades. In practically all modern electronic measurement systems the modifier will consist of one complex integrated circuit or a board containing many such circuits. The progress in this field was made possible by the development of the silicon planar technology. However, the further introduction of electronic systems into many applications is being hampered by the lack of suitable transducers whose inputs and outputs are compatible with, for instance, microprocessors. Therefore, it makes sense to apply silicon planar technology to the transducer field as well and to develop chips that are sensitive to light, pressure, displacement, humidity, pH, etc.

3 Silicon and its effects

Planar integrated-circuit technology would not have been possible without silicon and silicon dioxide. Silicon proved to be an element which could be purified to staggering levels and which shows a bandgap (1.1 eV) very suitable for the fabrication of semiconducting devices with a wide temperature range. If a wafer of silicon is heated in an atmosphere of

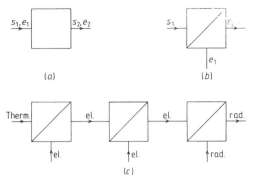

Figure 3 Graphic symbols for (a), a self-generating transducer; (b), a modulating transducer (s = signal, e = energy); and (c), an electronic thermometer with a temperature-sensitive transistor as input transducer and a liquid-crystal display as output transducer.

Table 1 Physical and chemical effects in silicon to be used in self-generating and modulating transducers.

Signal domain	Effects for selfgenerating transducers	Effects for modulating transducers
Radiant	Photovoltaic effect	Photoconductivity
Mechanical	—	Piezoresistance
Thermal	Seebeck effect	Temperature dependence of resistance
Magnetic	—	Magnetoresistance
Chemical	Galvano-electric effect	Ion-sensitive field effect

oxygen, a film of silicon dioxide forms on the surface. The film thickness can be accurately controlled by temperature and time. By photolithographic means openings can be easily, and with great precision etched, in the SiO_2 layer. When the wafer is exposed at high temperature to an n-type or p-type dopant containing gas mixture, the dopants will diffuse into the silicon only through the etched openings. The remaining SiO_2 layer prevents this diffusion in all other areas. This means that SiO_2 layers can be used as mask layers for the selective introduction of dopants. It might be useful to remark here that germanium and gallium–arsenide oxide layers do not form good masking layers, making them not very suitable for use in large-scale integration. Improved photolithographic techniques allowing structures down to 1 μm, epitaxy, ion implantation, metallisation, bonding and encapsulation techniques have all made the microelectronics revolution possible. When silicon planar technology is also used to produce sensors, it is necessary for certain physical and chemical effects to occur in silicon, which are suited to use in signal conversion.

The result of an overall analysis of the common effects occurring in silicon are shown in table 1. These effects are divided into those for self-generating transducers and those for modulating ones.

It appears that silicon is not suited to the construction of self-generating transducers in the mechanical and magnetic signal domains. The symmetry of the silicon lattice renders this material nonpiezoelectric. Therefore, it is not possible with silicon to generate a voltage by stressing a silicon crystal. Further, it is impossible with silicon to construct a device that operates without any auxiliary energy source and that can generate a voltage, when subjected to a magnetic field.

In order to make silicon micro-transducers that are sensitive to mechanical or magnetic signals, there are two possibilities. In the first one the piezo- and magnetoresistance effects are such that it is feasible to make modulating transducers for mechanical and magnetic signals. In the next section devices making use of such effects will be presented. The second possibility is to deposit layers on top of the silicon wafer, which show the desirable effects for the fabrication of self-generating transducers. In order to take advantage of the favourable features of the planar technology, it is then necessary for the extra deposition process to be compatible with the normal processing sequence and capable of being added without making too many drastic changes in the overall process. In the above two possibilities the silicon can be covered by a piezoelectric ZnO layer or a magnetic NiFe layer to make it suitable for either the mechanical or magnetic signal domain, respectively.

Examples of these devices will also be described in the next section and are considered to belong also to the silicon micro-transducer group.

Yet another reason to employ layers deposited by compatible technology is that silicon, though sensitive to signals of the relevant signal domain, might not show the desirable range of sensitivity. For instance, silicon, because of its bandgap of 1.1 eV and the wavelength dependence of its spectral absorption factor, is sensitive to wavelengths roughly between 0.4 and 1.1 μm. Deposition of a semiconducting layer with a smaller bandgap would extend the spectrum further in the infrared. Peripheral electronic circuits can then be integrated in the silicon substrate for further signal processing.

4 Review of the presently known silicon micro-transducers

The number of transducers whose fabrication is mainly based on the silicon planar technology with or without compatible processing steps and which are described in the scientific literature is already rather large. On the contrary, the number of devices which are commercially available is still strikingly small. This reflects the fact that only during the last few years has the importance of the field become recognised. Further, the development of reliable devices requires quite a few years of research and development. In the scientific world the effort needed to develop a successful transducer is often underestimated.

In this review an attempt is made to be as comprehensive as possible. This makes it necessary to give only very short descriptions of the devices. However, in the accompanying references a fuller account of the devices can be found.

There are different ways to group the silicon micro-transducers. It is possible to group them according to the main physical effects employed in the transducer or according to the signal domains to which the required information belongs. For instance, a proximity sensor can be constructed by using optical, mechanical, magnetic or even thermal physical effects. In this review, however, the latter applies, so that the proximity sensor will be described as a sensor belonging to the mechanical signal domain.

4.1 *SMT's for radiant signals*

Silicon can be used to construct sensors for a wide range of electromagnetic radiation wavelengths, from γ-rays to infrared. Sensors for visible light and the near infrared have especially been studied. Extensive literature on these devices exists (Pankove 1971, Ross 1979), and numerous products are also commercially available (Härtel 1978). In the field of radiation detection, silicon was already many years ago considered very suitable for sensor fabrication. All kinds of silicon devices have been invented, mainly for the measurement of light intensity. Devices are currently available for sensing light spots, line images and area images (Beynon and Lamb 1980). In this section the most important ones will be very briefly reviewed.

(i) *Photoconductors*

Photoconductors or photoresistors can be made in silicon with planar technology by creating n-type and p-type areas in the usual way. Incident radiation creates electron–hole pairs, causing a change in conductivity. A high sensitivity is obtained when the lifetime of the charge carriers is large compared to the transit time, i.e., the time required to move between the electrodes. However, this makes the device suitable for only low-frequency applications.

(ii) *Photodiodes*

Silicon photodiodes can be used with either reverse or forward bias. When a reverse bias is applied, incident light creates electron–hole pairs in the depletion region and its surroundings which are swept across the junction. This results in a photocurrent, which is a measure of the light intensity. To improve the frequency response of a photodiode, the device is fabricated in such a way that a nearly intrinsic layer is positioned between the p-type and n-type layers, which is completely depleted during operation. This structure has a low capacitance, whereas an electric field is present in the depletion layer for separating the charge carriers.

A photodiode can also make use of the photovoltaic effect. Here also, the depletion-layer field between the p-type and n-type layers will, again without a bias voltage, separate the electrons and the holes generated by the incident light. The n-type layer becomes negative with respect to the p-type layer. A photodiode of this type, which is in fact a solar cell, can be described by a p–n junction with a light-intensity dependent current source in parallel with the junction. Very fast photodetectors can be made with Schottky barrier diodes, consisting of rectifying metal-to-silicon contacts. Because only majority carriers are involved and the surface barrier can be adjusted by choosing the proper metal, very fast photodetectors with adjustable spectral sensitivity can be made. Photodiodes that are operated at such a high reverse-bias voltage that avalanche multiplication takes place are called avalanche photodiodes. When a Schottky barrier is also employed, very high-speed photodetectors can be made with high sensitivity, even to UV light.

(iii) *Phototransistors*

From the fact that p–n junctions are very suitable for photodetectors it is not difficult to imagine that transistors can also be used to this purpose. In a properly designed phototransistor the emitter is made very small, so that light can reach the base-collector junction. When the base is not connected the photocurrent generated in this junction will be multiplied by the current gain of the transistor.

When an npn phototransistor and a pnp transistor are properly connected together a photothyristor circuit results. Sufficient illumination of the photothyristor triggers the circuit.

(iv) *Charge-coupled devices* (CCD's)

When the capacitance-voltage curves of MOS (Metal-SiO$_2$-Si) diodes are measured under illumination and p-type Si and positive voltages on the metal electrode are used, the total capacitance increases with increasing light intensity. This is caused by the generation of electron–hole pairs whose electrons gather at the Si–SiO$_2$ interface.

The capacitance of the illuminated MOS diode is then equal to the capacitance of the SiO$_2$ layer. The depletion-layer contribution to the capacitance is in that case very small. It is not difficult to imagine, that these effects can be used for the design of photo-field-effect transistors. In such transistors the channel conductivity can be modulated by incident light.

Photo-MOS transistors have not become as popular as bipolar photodetectors for measuring light-spot intensity. However, when MOS diodes are connected to each other in such a way that a shift register results, the devices have been shown to be very useful for the sensing of line and area images (Beynon and Lamb 1980). A line image projected on the CCD register generates charge packets beneath the gate electrodes, which can be shifted in either direction by manipulating the gate voltages. When a large number of such CCD registers are positioned parallel to each other, two-dimensional images can also be sensed. CCD arrays with more than 10^6 picture elements have been processed.

One problem with such a CCD image sensor is blooming: when a bright spot occurs in the image, there is a tendency for excess carriers to spread to neighbouring picture elements. Several methods have been found to reduce this problem. In one CCD, called a junction CCD, charge transfer in an epitaxial layer is initiated by p–n junctions rather than by gate electrodes. Excess charges can easily be removed via the junctions (Hartgring and Kleefstra 1978). The work on CCD imaging is currently attracting continuously increasing interest, because it might open the way to low-cost solid-state TV cameras.

Light intensity can also be measured by observing the temperature of a small black body that is heated because of the absorbed incident radiant energy (bolometre principle). As we will describe later on, very suitable temperature sensors can be designed in silicon, so that this method deserves future attention.

In the beginning of this section we stated that planar silicon devices can also be used beyond the visible light range. However, not much work has been reported in this field. There is only one known silicon detector which is made with planar technology on high-resistivity silicon and which is capable of detecting both α and β particles and x-rays (Kemmer 1980). Nevertheless, it can be expected, that the advances in silicon planar technology in the near future will make possible devices with improved resolution with respect to energy and beam position.

4.2 SMT's for mechanical signals

Unlike the radiant-signal field, silicon has up to now rarely been used for commercially available sensors in the mechanical-signal domain. Though the scientific literature already contains discussions of a number of rather interesting devices, it can be expected that in the near future this signal domain will be the focus of more intense R&D efforts. The most important devices currently being made are to measure pressure or weight, but a number of silicon devices have also been developed to sense acceleration force, position and air flow.

4.2.1 *Pressure transducers.* The best-known device is the so-called diffused silicon strain-gauge diaphragm, first described in 1969 by Gieles (Gieles and Somers 1973). This device is based on the piezoresistive effect. Normal metal strain gauges have a gauge factor (relative change in resistance divided by relative change in length) of around two. Semiconductors, however, show a gauge factor of around 200. This large value is due to the fact that an anisotropic strain changes the average mobility of the charge carriers (Mason 1969). The most simple devices making use of this effect consists of a diffused resistor in a silicon substrate (Angell 1978). More sophisticated devices as the ones designed by Gieles, consist of a very thin (15 μm), etched n-type silicon pressure diaphragm with a diameter of 1 mm in which a Wheatstone bridge of four p-type doped resistors has been diffused. Based on this device but using somewhat larger diaphragm diameters, several industries now offer a whole series of pressure transducers in a pressure range from 0–0.6 kPa to 0–35 kPa (National Semiconductor 1977, Honeywell 1978). A disadvantage of these silicon pressure transducers is their high temperature sensitivity. By adding a simple three-resistor network (Castle 1974) or by using epitaxially grown resistors on a sapphire diaphragm (Allan 1979) it is possible to compensate for the temperature sensitivity and to extend the temperature range of operation.

A pressure transducer using polycrystalline silicon as diaphragm material is also described in literature (Jaffe 1974). The piezoresistor is fabricated by ion implantation directly in the polycrystalline silicon.

In the above-described silicon diaphragm pressure transducers the piezoresistance effect is employed to convert strain into an electrical signal. It is also possible to make a silicon device in which a capacitance change is detected and used as a measure of the pressure (Sander *et al* 1980). In this case, the capacitance is formed between a metallised glass plate and a thin silicon diaphragm separated by a 2 μm gap. The device is 20 times more sensitive than comparable piezoresistive transducers and requires less power. When pressure is applied on the emitter surface of a planar diffused transistor by means of a stylus, the characteristics of the transistor will change. The emitter-base voltage increases with increasing pressure while the gain decreases (Bulthuis 1965). The effects are explained in terms of a change in the band gap. Based on this effect a silicon pressure transducer was made, in which the signal processing circuits and the sensor were integrated on the same 2×2 mm chip. This results in an output of about 2 mV g^{-1} (Veen 1979). MOSFET's and JFET's also appear to be pressure sensitive (Nuzillat and Helioui 1973, Wlodarski 1973). By integrating suitable combinations of devices it is possible to make transducers that have ten times greater sensitivity than diffused piezoresistor gauges and better thermal stability.

The sensitivity of a MOSFET to strain can be improved when, by means of a compatible technology, piezoelectric ZnO or CdS layers are deposited between the gate metal and the gate oxide layer (Greeneich and Muller 1972).

Applying pressure to the structure changes the polarisation of the piezoelectric layer, which can then be electrically detected as a change in the drain current. Gauge factors in excess of 9000 have been measured. In order to make the piezo-FETS suitable for detecting high-frequency surface-acoustic waves the ZnO and diffused MOS (DMOS) technologies were combined, yielding a device capable of detecting 28 MHz surface-acoustic waves (Yeh and Muller 1976).

4.2.2 Acceleration transducer.

Another important mechanical signal is acceleration. Several silicon devices for the measurement of this parameter are described in the literature.

One device consists of a 200 μm-thick silicon mass of around 1 mm^2 serving as an inertial reference which is connected to a 15 μm-thick and 0.5 mm long cantilevered beam of silicon (Roylance and Angell 1979). A p-type piezoresistor is diffused in the beam. Acceleration along a direction normal to the plane of the mass causes the thin beam to bend, and a change in the piezoresistor can then be electrically detected.

The entire structure is fabricated by a sequence of very sophisticated processing steps of which the KOH anisotropic etch steps are most notable. The device can measure acceleration from 0.1–1000 ms^{-2}. Another device, also batch-fabricated in silicon by a number of innovative etching steps, covers an area of only 0.015 mm^2 (Petersen and Shartel 1980). Conversion of the mechanical signal into an electrical one is performed by measuring capacitance changes of a metal-coated SiO$_2$ cantilever beam with respect to the substrate. A sensitivity of 0.2 mV/ms^{-2} is achieved. Yet another device uses a beam consisting of a sandwich of silicon, silicon dioxide, piezoelectric zinc oxide, metal and silicon dioxide (Chen et al 1980). The polarisation of the ZnO layer is detected by a depletion-type, p-channel MOS transistor. The measurement range is from 1–10^5 ms^{-2} with a sensitivity of 4.7 μV/ms^{-2}.

4.2.3 Position transducers.

By using a light beam and a mirror attached to the object whose position is to be detected, object position can be translated into light-beam position. With a specially designed silicon photodetector it is possible to obtain X- and Y-signals exactly proportional to the coordinates of the light spot, where the light beam is falling upon the detector (Noorlag and Middelhoek 1979). The photoconductor consists of an 8 × 6 mm^2 pn diode reverse-biassed at about 10 V. The upper p-type layer has contacts at the edges parallel to the Y-axis, while those of the lower n-type layer are parallel to the X-axis. The photocurrent distribution amongst the contacts depends on the distance between the light spot and the contacts. Linearity is better than 0.5% full scale and the resolution is about 5 μm. Larger devices (15 × 15 mm^2) have also been produced.

Another method for measuring position, motion, displacement, etc., is based on the use of a magnet with a matching magnetic-field-sensitive sensor. Because the measured field depends on the distance between sensor and magnet it is possible to convert this mechanical signal into an electrical signal. Devices made in silicon for the measurement of magnetic fields will be discussed in one of the next sections.

4.2.4 Flow meters.

A very important physical quantity in the mechanical-signal domain is gas or liquid flow. Devices presently used for this purpose are bulky and inaccurate and seldom give suitable electronic output signals. Two flow sensors using silicon are described in the literature. One device makes use of a p-type Wheatstone bridge diffused into an n-type substrate (Van Putten and Middelhoek 1974). When an air flow passes along the chip parallel to two resistors, the resistors normal to the flow are cooled to a greater extent than the two other resistors. The bridge unbalance is then a measure of the flow velocity. In another device three transistors are placed on a silicon chip (Huysing 1980). The transistor in the centre is used as a heating device, while the ones to the left and right serve as temperature sensors. When an air flow is present, a temperature difference between the two transistors is measured, due to different laminar boundary layer thicknesses. This temperature difference is proportional to the square root of the flow velocity. Adding peripheral electronic circuits leads to a sensor with a sensitivity of 10 mV/ms^{-1}. Accuracies better than 1% are feasible.

4.3 SMT's for thermal signals

All semiconductors and thus also silicon electron devices show temperature dependences which are usually considered to be very undesirable. However, an extensive literature illustrates that this property can also be put to use to develop sensors that can accurately measure the temperature. To that end, thermocouples, resistors, diodes, transistors and IC's are described.

4.3.1 Thermocouples.

In practice, temperatures are very often measured by means of thermocouples. When two dissimilar conductors are joined together and the junctions are at different temperatures, the resulting voltage is a measure for the temperature difference. As no auxiliary power source is required a thermocouple is a self-generating transducer. By means of thermocouples consisting of evaporated aluminium films and diffused p-type and n-type layers it has been shown that sensitivities up to 1.5 mV K^{-1} are possible (Kerkhoff and Meijer 1979). By using compatible technology an infrared detector with thermocouples can also be fabricated (Lahiji and Wise 1980). Sixty bismuth–antimony thermocouples connected in series are deposited on a silicon substrate. Through anisotropic etching of the silicon beneath the hot junctions, which are exposed to the infrared radiation, the thickness of the silicon is reduced to 1 μm. This is necessary to obtain a high thermal resistance between the hot and cold junctions. The detector shows a responsivity of 7 V W^{-1} with a time constant of 15 ms.

4.3.2 Resistors.

Since the density and mobility of charge carriers in semiconductors depend on the temperature, it is possible to use the resistivity for the measurement of the temperature. In small bandgap materials at room temperature the temperature behaviour is like that of an intrinsic semi-conductor. Sensors made with such materials are called thermistors and are very sensitive to temperature. Silicon at room temperature behaves like an extrinsic semiconductor and shows only moderate temperature sensitivity because only the carrier mobility, not the carrier density, is temperature-dependent. Nonetheless, a few silicon temperature sensors are commercially available.

Using phosphorous-doped silicon layers with a very high impurity concentration of $3 \times 10^{24}\,\mathrm{m^{-3}}$ suitable resistance thermometers can be made for the temperature range around 4 K (Kirsch et al 1972). When highly accurate temperature sensors are required it is possible to deposit metal, e.g. platinum, films on top of a silicon dioxide-covered silicon substrate. The resistance of the metal film is then used for the measurement of the temperature. Furthermore, the silicon substrate can be used for signal-processing purposes.

4.3.3 Diodes and transistors.

When a diode is forward-biassed it has been shown that in an ideal diode the voltage across the junction almost linearly decreases with increasing temperature, at least when the current through the diode is kept constant (McNamara 1962).

A number of diode thermometers based on this effect has been proposed. Because diodes do not behave as ideally as Shockley's diode theory assumes, the linearity is often not sufficient and reproducible enough to make high-accuracy thermometers. In p–n junctions generation-recombination in the depletion layer and recombination at the semiconductor surface give rise to a current contribution that is not accounted for in Shockley's diode theory. Therefore, many authors have suggested using a transistor and its more ideal V_{BE}–I_C characteristics (Davis and Coates 1977, Verster 1972). For constant I_C such devices show a sensitivity of about 2 mV °C^{-1}. By adding signal-processing circuits on to the chip it is possible to control the operating point in such a way that the linearity and stability are further improved. A circuit which enables the mass production of thermometers with ±0.1°C linearity over the −50 to +125°C temperature range is described in literature.

Interchanging the sensor requires only a one-point temperature calibration. A more precise circuit requires a three-point calibration but yields ±0.01°C accuracy (Ohte and Yamagata 1977).

4.3.4 Dual transistors and IC transducers.

When two bipolar transistors are operated at a constant ratio of the emitter-current densities, the difference ΔV_{BE} in the base-emitter voltages is proportional to the absolute temperature (PTAT). Different emitter-current densities can be achieved in two ways. To begin with, two identical transistors operated at two different collector currents, held at a constant ratio, can be used. However, it is more convenient to make two transistors with unequal emitter areas at equal collector currents (Timko

1976). When, for instance, a ratio of 8 is used the sensitivity is 0.1792 mV K^{-1}.

An integrated temperature transducer results from integrating sophisticated electronic circuitry together with the dual transistor. By laser-trimming several resistors on the chip, a transducer with an overall accuracy of ±0.5°C between −75°C to +150°C is achieved (Timko 1976). In many applications, an output signal in Fahrenheit, Celcius or another scale is required. This can be obtained by using the above calibrated PTAT temperature transducer, a calibrated voltage reference, and a calibrated resistor network for the scale factor. In the past, three non-independent calibration steps have been required, adding to cost and complexity. Recently, a new IC temperature transducer was described in which the intrinsic bandgap voltage is used as a reference and whereby only simple calibration is needed (de Haan and Meijer 1980). The output is zero at 0°C, and the sensitivity of the sensor version used in a clinical thermometer is 100 mV °C^{-1} in the temperature range between 20°C and 50°C. The sensor shows in this range an absolute error of ±0.1°C and a long-term stability of 0.03°C.

4.4 SMT's for magnetic signals

Because silicon is not magnetic it is impossible to affect the motion of charge-carriers in silicon by changing the magnetisation direction by applying an external magnetic field. Furthermore, a self-generating transducer is hardly possible with silicon. Fortunately, silicon lends itself rather well to the construction of modulating transducers, and many such devices are presented in literature. These devices are based on the Hall effect or the magnetoresistance effect and encompass Hall plates, magnetoresistors, diodes, transistors, dual transistors, MOSFET's and carrier-domain devices.

The Hall effect and magnetoresistance originate from the fact that a charge carrier moving in a magnetic field encounters the Lorentz force, which is mutually perpendicular to the directions of the charge carrier velocity and the magnetic field. Under the influence of this Lorentz force the charge carriers will 'pile up' at one end of the conductor, so that charge depletion will occur at the other end. This gives rise to an electrical force equal but opposite in direction to the Lorentz force. The Hall effect is that effect in which a voltage is measured perpendicular to the current direction. When the Hall voltage is short-circuited the electric field is reduced to zero and the charge carriers will be deflected by the Lorentz force. This results in longer pathlengths for the charge carriers and thus in an increased resistance. This effect is known as the geometric magnetoresistance effect. Both Hall and magnetoresistance effects are used for making magnetic-field-sensitive SMT's.

4.4.1 Hall plates and magnetoresistors.

From a simple analysis it can be deduced that the Hall voltage is proportional to the current through the plate and to the magnetic-field amplitude.

Moreover, the Hall voltage is inversely proportional to the charge-carrier density and the thickness of the plate. When planar silicon technology is applied, Hall plates can be successfully made using thin, highly resistive n-type layers on a p-type substrate. The contacts for the current and the Hall voltage are obtained by masked n^+ diffusions.

Its high Hall coefficient makes silicon very suitable as a material for making Hall plates. Mounting an integrated Hall plate on to a nonmagnetic package produces a compact transducer. Great care must be taken in bonding the device into the package, since mechanical strain tends to cause uncontrollable offsets of the Hall voltage. The silicon Hall plate in combination with some signal-processing circuits in the same chip can be used for the measurement of magnetic fields or as a magnetic-field-sensitive switch. Commercially available analogue Hall plates have sensitivities of around 100 mV mT^{-1} with offsets smaller than 35 mT, which can be compensated by external circuitry (Lachmann 1979). Good linearity is obtained up to fields of 100 mT.

The magnetoresistive effect in semiconductors is proportional to the square of the charge-carrier mobility and the square of the magnetic field. As a consequence, semiconductors with high mobility are preferred. InSb with an electron mobility of 7.8 m²/V s is often used. Si is less suitable because its electron mobility is only 0.14 m²/V s.

When for some reason silicon is very much favoured for making transducers based on the magnetoresistance effect compatible technology can provide the solution. For instance, one can imagine combining the high-mobility material InSb with signal-processing circuits on one silicon chip. However, in practice another approach is followed. Because of the interaction between moving charge carriers and the magnetisation in Permalloy (80% Ni and 20% Fe) films the resistance of such films change when the magnetisation rotates in the plane of the film. The change is largest when current and magnetisation form a 45° angle. Depending on the deposition parameters the external magnetic field necessary to cause such a magnetisation rotation can be made very small. To obtain a 45° angle between current and magnetisation a small permanent magnet is required. A much more attractive method is known as the 'barber pole' configuration (Kuijk et al 1975). By evaporating highly conductive gold strips on top of the NiFe layers, which are at an angle of 45° with the magnetisation direction, the electric field can also be made to form a 45° angle with the magnetization. Maximum resistance changes of 2% are possible, whereas the smallest detectable fields are around 10^{-7} T.

The above-described structures are on deposited silicon substrates, which can also contain signal-processing circuits. The sensor is used for the fabrication of a magnetic recording head (Druyvesteyn 1980).

4.4.2 *Magnetodiodes.* Because of the Lorentz force charge carriers in semiconductors are diverted from their straight path by a magnetic field. A diode can be constructed in such a way that the charge carriers are diverted towards a surface with high surface recombination. The I–V characteristics of the so-called magnetodiode consequently exhibit a dependence on the magnetic field.

In order to obtain a large magnetic-field sensitivity the difference between low and high recombination rates must be as large as possible. Diodes made in silicon-on-sapphire technology especially have been shown to be very suitable as the recombination dissymmetry of the Si–SiO$_2$ and Si–Al$_2$O$_3$ surfaces is rather distinct (Lilienkamp and Pfleiderer 1977).

4.4.3 *Magnetotransistors.* Clearly, also transistors incorporating the just described effects occurring in magnetodiodes can be made. However, a more interesting way of making a magnetotransistor is based on structures consisting of one emitter, one base and two collector contacts. Such structure have been realised with different techniques. One of the oldest devices, invented by Hudson, is made up of a transistor in which the collector is obtained by diffusion. This collector consists of two parts which are symmetrical (Cushman 1969). Without a magnetic field the currents to the collectors are equal. When a magnetic field is applied in the plane of the silicon wafer and perpendicular to the imaginary line connecting the collectors, more charge carriers are directed to one of the collectors at the expense of the other collector. The current difference can be measured and is a linear function of the magnetic field. When the magnetic field is rotated over 90°, the effect is reduced to zero. A similar device consists of a chip, with the collector contacts at the bottom, separated by a groove (Flynn 1970). The collector-base voltage has to be large enough so as to extend the depletion layer from the base to the groove. A magnetic field applied parallel to the groove causes the collector currents to become unequal, which then can be measured.

In still another approach a field aided lateral pnp structure with two collectors is employed (Davies and Well 1971, Halbo and Haraldsen 1980). Minority carriers are injected by the emitter and collected at the collectors, which are symmetrically positioned at a short distance from eath other. The complete lateral transistor is positioned between two ohmic contacts, which allows a lateral drift field to be generated to accelerate the injected charge carriers. A magnetic field normal to the substrate will deflect the charge carriers, resulting in unequal collector currents. In another version of the lateral magnetotransistor the elongated collector contacts are parallel to the emitter contact, but at opposite sides of the emitter. Such a magnetotransistor is, like the others, sensitive to fields parallel to the substrate and the emitter (Mitnikova et al 1978).

Recently a magnetotransistor made by normal bipolar technology and employing a split buried layer to increase the sensitivity was presented. Two collector contacts are positioned at opposite sides of the emitter. The buried layers between the collector layer and the substrate extends from beneath the collector contacts to just beneath the edges of the emitter-base structure (Zieren 1980). These devices appear to have a linearity better than 0.3% in the range 10^{-4}–1 T. A four-collector structure based on the same principle has also been developed. It is capable of simultaneously measuring the magnitude and the direction of a magnetic-field vector. Another magnetic-field-sensitive two-transistor structure makes use of the Hall effect. The transducer consists of an integrated Hall plate and two transistors arranged so that the Hall plate is used as the common base of the transistors. The Hall base region has two contacts. When a Hall current flows between them, a Hall voltage appears in such a way that one transistor receives an increased base voltage and the other a decreased one. The resulting difference in collector currents can be measured and is a measure of the magnetic field (Takamiya and Fujikawa 1972). In comparison to the other collector devices based on Lorentz force deviation, this device shows a less linear response.

Just like transistors made with bipolar technology, transistors made with MOS technology also show magnetic-field sensitivity. In such MOSFET's the channel between the source and drain contacts is used as a Hall plate. Under the gate in the centre two diffused contacts are fabricated for measuring the Hall voltage (Callagher and Corak 1966). Bear in mind, as was stated at the beginning of this section, that the Hall voltage is inversely proportional to the thickness of the Hall plate. Thus, the channel under the gate can be reduced by the gate voltage and by the drain voltage. The highest Hall voltages are obtained when the device is operated near pinch-off. In another device described in literature the Hall contacts are positioned closer to the pinch-off region near the drain so that still higher output voltages could be detected (Fry and Hoey 1969).

Lastly a magnetic-field-sensitive device that deviates very much from the above-described structures is to be presented. It is known, that under certain circumstances in pnpn structures current is not homogeneous but is localised in a small region known as the filament or carrier domain. When a magnetic field is applied normal to the filament, the filament can be displaced due to Lorentz forces (Persky and Bartelink 1974). In a so-called carrier-domain magnetometer the structure is circular and the magnetic field causes the carrier domain to rotate with a frequency which is a function of the magnetic field (Gilbert 1977, Manley and Bloodworth 1978).

4.5 SMT's for chemical signals

The growing interest in better biomedical instrumentation, chemical process and environmental control has increased the demand for accurate sensors for chemical measurands. This demand has spurred many laboratories on to research into silicon devices for this purpose. Silicon devices are very sensitive to trace impurities and contamination, so that the semiconductor industry has taken great pains to develop passivation and encapsulation techniques. When silicon devices are to be used for chemical sensors the encapsulation problem has to be solved.

The literature on this subject is already very extensive (Bergveld and Zemel 1981, Cheung et al 1978, Bergveld and De Rooij 1980). In this section only the most important transducers will be presented.

4.5.1 *The ISFET.* The first chemical SMT, labelled an ISFET (ion-sensitive FET), was presented as early as 1968 (Bergveld 1970). The structure consists of a conventional MOSFET, except that no gate electrode is used, and further, the thermally grown gate oxide is covered with an ion-selective layer of Si_3N_4, Al_2O_3, polymer. When these devices are immersed in an electrolyte, the characteristics of the ISFET are affected by the ionic activity of the electrolyte. This makes it possible to produce such devices as pH sensors. Though various experimental data have been offered up to now, agreement about the exact functioning of the device has still not been reached (Siu and Cobbold 1979). The need for a reference electrode is also disputed. In spite of the controversies, the ISFET's have proved their usefulness as disposable sensors for biomedical applications.

4.5.2 *The palladium-gate MOSFET.* When the conventional aluminium gate of a MOSFET is replaced by a palladium gate the MOSFET characteristics become sensitive to hydrogen (Lundstrom et al 1975). It is assumed that the adsorption of hydrogen on the palladium surface and subsequent diffusion of atomic hydrogen to the palladium–oxide interface causes changes in the palladium silicon work function. The threshold voltage of the device is then a measure for the hydrogen concentration. Sensitivities down to 10 PPM H_2 are feasible. Moreover, it has been shown that the device is also sensitive to H_2S and NH_3.

4.5.3 *Polymer covered silicon devices.* A large number of polymers are known of which the resistivity and the dielectric constant change when they are exposed to gasses. When such polymers are deposited on top of interdigital electrodes or MOSFET's in silicon, transducers can be made that respond to a number of gasses, depending on the polymer used (Senturia 1975). To measure the resistivity as a function of the gas content a new device, the so-called charge-flow transistor (CFT), was later developed (Senturia et al 1977). These sensors are sensitive to CO, CO_2, CH_4, SO_2, NH_3 and many other gasses. Because the devices are also very sensitive to H_2O, it was further possible to construct a humidity sensor.

4.5.5 *Humidity sensors.* Most of the present humidity sensors are not distinguished by a high accuracy, a wide temperature and humidity range of operation and low cost. Moreover, they do not supply microprocessor-compatible output. However, several silicon-based solutions to the problem can be presented.

First, as already described, the resistance of a suitable polymer layer as a function of the humidity can be measured with a CFT (Senturia *et al* 1977). In another structure the capacitance of a meander-like structure on top of a silicon substrate is measured as a function of the relative humidity (Jachowicz 1980).

In a third sensor a silicon chip consists of an interdigital electrode structure for detecting capacitance changes due to dew formation on the chip and a transistor for measuring the chip temperature (Regtien 1980). The chip is mounted on a miniature Peltier element, used as the cooling device. During operation, the temperature of the chip is lowered until water vapour from the air starts to condense on the chip, thus causing a capacitance change. The temperature at which condensation starts is maintained by means of a control circuit, so that the dew coverage of the capacitor structure remains the same. The temperature is then a measure of the relative humidity.

Based on this principle a very small, accurate humidity sensor is constructed, showing a nonlinearity less than 0.1 K over a temperature range of 50 K.

4.5.6 *Oxygen sensor.*

Silicon planar technology is very suited to mass production. This explains why silicon is often even used for devices in which the semiconducting properties of silicon are not used. One such device is the oxygen sensor (Honig 1980).

Spiral channels are etched on a silicon substrate. These serve as a chamber for an electrolyte consisting of an aqueous solution of NaCl. Two silver electrodes are deposited for use as anode and cathode. When a voltage is applied, the current is a measure for the oxygen pressure in the cell. The spiral channel is covered by a membrane through which oxygen can diffuse. When the membrane is properly chosen the current is also a measure for the outside oxygen pressure. Because Ag and Cl are used when the cell is in operation, the lifetime is limited by the amounts of Ag and Cl present.

5 Conclusions and future outlook

The penetration of microelectronics into traditionally non-electronic markets has been seriously hampered by the lack of low-cost, accurate transducers, which can generate an electrical output compatible with microprocessors or other electronic circuits. However, the use of silicon planar technology might bring forth a new generation of silicon microtransducers (SMT's). When silicon itself does not show the required physical effects, other layers can be used which are deposited by compatible technology on the silicon substrates. Because the silicon technology lends itself so well to mass production, it will also increasingly be used in the future as a fabrication method to make devices not requiring the semiconducting nature of silicon.

SMT's already have a long tradition in the radiant-signal field. Many devices are commercially available, and greater sophistication can be expected. In particular, signal-processing circuits will be added to the sensor, on the same chip, creating so-called 'smart sensors'. It is to be expected that new sensors for x-rays or nuclear radiation will be developed in the near future. SMT's for measuring mechanical sensors have not yet reached such acceptance. Only the diffused pressure transducer is widely in use. A large number of interesting and innovative sensors will undoubtedly appear in the future. However, it is at present still a largely underdeveloped field.

On the other hand, SMT's for measuring thermal signals are far along in development. Much research was initially done on the temperature sensitivity of silicon devices, making it easy to put this acquired knowledge to good use. In the future new IC temperature sensors will be developed which require less and easier calibration and which are usable in a wider temperature range. SMT's for measuring magnetic signals are still lagging behind, plagued by low sensitivity and offset problems. Good sensors for the field range below 0.1 mT are not yet available, although most probably the addition of signal-processing circuits on the chip will lead to successful devices in this range.

SMT's for chemical signals suffer from the encapsulation problem. A sensor must be reliable, reproducible and have a long lifetime. However, the need for new sensors in the chemical-processing field is so pressing that the enormous efforts being expended in this area will very likely soon lead to acceptable solutions.

SMT's promise a fruitful market for the semiconductor and instrumentation industries. It can be assumed that the semiconductor industry will focus on low-cost, versatile devices, which can be mass produced and which also have a mass market.

The smaller instrument companies can then focus on the higher-priced specialties for which many believe a growing market will also develop. Thus, as the foregoing indicates, the future outlook for SMT's with regard to research possibilities, and marketability is clearly very promising.

References

Allan R 1979 Transducer sports SOS diaphragm
Electronics November **22** 42–4

Angell J B 1978 Micromachined silicon transducers for measuring force, pressure and motion
Proc. ESSCIRC 33–6

Bergveld P 1970 Developments of an ion-sensitive solid-state device for neuro-physiological measurements
IEEE Trans. Biomed. Eng. **BME-17** 70–1

Bergveld P and De Rooij N 1980 The development of chemical sensitive electronic devices
Solid state sensors (Deventer: Kluwer) 119–39

Bergveld P and Zemel J N (ed) 1981 *Chemically sensitive devices* (Lausanne: Elsevier Sequoia)

Beynon J D E and Lamb D R 1980 *Charge-coupled devices and their applications* (London: McGraw-Hill)

Bulthuis K 1965 The effect of local pressure on silicon p–n junctions
Philips Res. Rep. **20** 415–31

Callagher R C and Corak W S 1966 A metal-oxide semiconductor (MOS) Hall element
Solid-State Electron **9** 571–80

Castle P F 1974 A temperature compensated silicon strain transducer
Strain **10** 22–5

Chen P, Jolly R, Halac G, Muller R S and White R M 1980 A planar-processed Pi-FET accelerometer
Proc. IEDM Washington 848–9

Cheung P W, Fleming D G, Ko W H and Neuman M R 1978 (ed). *Theory, design and biomedical applications of solid-state chemical sensors* (West Palm Beach: CRC Press)

Cushman R H 1969 Transistor responds to magnetic fields
EDN February 15 73–8

Davies and Wells 1971

Davis C E and Coates P B 1977 Linearization of silicon junction characteristics for temperature measurement
J. Phys. E: Sci. Instrum. **10** 613–6

Druyvesteyn W F 1980 Magnetoresistive sensor
Solid state sensors (Deventer: Kluwer) 87–94

Flynn J B 1970 Silicon depletion layer magnetometer
J. Appl. Phys. **41** 2750–1

Fry P W and Hoey S J 1969 A silicon MOS magnetic field transducer of high sensitivity
IEEE Trans. Electron. Dev. **ED-16** 35–9

Gieles A and Somers G 1973 Miniature pressure transducers with a silicon diaphragm
Philips Techn. Rev. **33** 14–20

Gilbert 1977

Greeneich E W and Muller R S 1972 Acoustic-wave detection via a piezo-electric field-effect transistor
Appl. Phys. Lett. **20** 156–8

Haan G de and Meijer G C M 1980 An accurate small range IC temperature transducer
IEEE J. Solid State Circ. **SC-15** 1089–91

Halbo L and Haraldsen J 1980 The magnetic field sensitive transistor: a new sensor for crankshaft angle position
Proc. SAE Congress Detroit February, 25–9

Härtel V 1978 *Optoelectronics* (New York: McGraw-Hill)

Hartgring C D and Kleefstra M 1978 Quantum efficiency and blooming suppression in junction charge-coupled devices
IEEE J. Solid-State Circ. **SC-13** 728–30

Honeywell 1978 *Diffused silicon transmitters*

Honig E P 1980 A miniature oxygen sensor
Solid state sensors (Deventer: Kluwer) 141–5

Huysing J H 1980 Monolithic flow sensors
Solid state sensors (Deventer: Kluwer) 39–48

Jachowicz R S 1980 MOS type miniature capacitance sensor for measuring humidity of gases
Proc. Eurocon, Stuttgart 654–6

Jaffe J M 1974 Monolithic polycrystalline silicon pressure transducer
Electron. letters **10** 420–1

Kemmer J 1980 Fabrication of low noise silicon radiation detectors by the planar process
Nucl. Instrum. Methods **169** 499–502

Kerkhoff H G and Meijer G C M 1979 An integrated electrothermal amplitude detector using the Seebeck effect
Digest Techn. Papers ESSCIRC, Southampton 31–3

Kirsch H C, Bachmann R and Geballe T H 1972 Silicon low temperature thermal elements
Temperature, its measurement and control in science and industry **4** (Pittsburgh: Instrument Society of America) 843–6

Kuijk K E ,Van Gestel W J and Gorter F W 1975 The barber pole, a linear magnetoresistive head
IEEE Trans. Mag. **11** 1215–7

Lachmann U 1979 SAS 231, an integrated Hall effect circuit with analog output
Components Report (Siemens) **14** 225–7

Lahiji G R and Wise K D 1980 A monolithic thermopile detector fabricated using integrated-circuit technology
Proc. IEDM, Washington 676–9

Lilienkamp P and Pfeiderer H 1977 An ESFI SOS magnetodiode
Phys. Status Solidi **43** 479–86

Lion K S 1969 Transducers: Problems and prospects
IEEE Trans. Industr. Electron. Contr. Instrum. **IECI-16** 2–5

Lundstrom K I, Shivaraman M S and Svensson C M 1975 A hydrogen-sensitive Pd-gate MOS transducer
J. Appl. Phys. **46** 3876–81

McNamara A G 1962 Semiconductor diodes and transistors as electrical thermometers
Rev. Sci. Instrum. **33** 330–2

Manley M H and Bloodworth G G 1978 The carrier-domain magnetometer: a novel silicon magnetic field sensor
IEE J. Solid-State and Elec. Dev. **2** 176-84

Mason W P 1969 Use of solid state transducers in mechanics and acoustics
J. Aud. Eng. Soc. **17** 506–11

Middelhoek S, Angell J B and Noorlag D J W 1980

Microprocessors get integrated sensors
IEEE Spectrum **17** February 42–6

Middelhoek S and Noorlag D J W 1980 The present and future of silicon micro-transducers (SMT's)
Proc. des Journées d'Electronique et de Microtechnique, Lausanne 57–73

Middelhoek S and Noorlag D J W 1981 Three-dimensional representation of solid state transducers
Sensors and Actuators 29–41

Mitnikova I M, Perisyanov T V, Rekalova G I and Shtynbner G 1978 Investigation of the characteristics of silicon lateral magnetotransistors with two measuring collectors
Sov. Phys. Semicond. **12** 26–8

National Semiconductor 1977 *Pressure transducer handbook*

Noorlag D J W and Middelhoek S 1979 Two-dimensional position-sensitive photodetector with high linearity made with standard IC technology
IEE J. Solid-State and Elec. Dev. **3** 75–82

Nuzillat G and Helioui H 1973 Transducteurs piézo-FET analogiques
Revue Techn. Thomson-CSF **5** 49–80

Ohte A and Yamagata M 1977 A precision silicon transistors thermometer
IEEE Trans. Instrum. and Meas. **IM-26** 335–41

Pankove J I 1971 *Optical processes in semiconductors* (Englewood Cliffs: Prentice-Hall)

Persky G and Bartelink D J 1974 Controlled current filaments in PNIPN structures with application to magnetic field detections
Bell Syst. Tech. J. **53** 467–502

Petersen K and Shartel A 1980 Micromechanical accelerometer integrated with MOS detection circuitry
Proc. IEDM Washington 673–5

Regtien P P L 1980 An integrated humidity sensor
Solid state sensors (Deventer: Kluwer) 109–17

Ross D A 1979 *Optoelectronic devices and optical imaging techniques* (London: Macmillan)

Roylance L M and Angell J B 1979 A batch-fabricated silicon accelerometer

IEEE Trans. Electron. Dev. **ED-26** 1911–7

Sander C S, Knutti J W and Meindl J D 1980 A monolithic capacitive pressure sensor with pulse-period output
IEEE Trans. Electron. Dev. **ED-27** 927–30

Senturia S D 1975 Fabrication and evaluation of polymeric early-warning fire-alarm devices
NASA CR-134764

Senturia S D, Sechen C M and Wishnewsky J A 1977 The charge-flow transistors: a new MOS device
Appl. Phys. Lett. **30** 106–8

Siu W M and Cobbold R S C 1979 Basis properties of the electrolyte SiO_2–Si system: physical and theoretical aspects
IEEE Trans. Electron. Dev. **ED-26** 1805–15

Takamiya S and Fujikawa K 1972 Differential amplification magnetic sensor
IEEE Trans. Electron. Dev. **ED-19** 1085–90

Timko M P 1976 A two terminal IC temperature transducer
IEEE J. Solid State Circ. **SC-11** 784–8

Van Dijck J G R 1964 *The physical basis of electronics* (Eindhoven: Centrex) 31–44

Van Putten A F P and Middelhoek S 1974 Integrated silicon anemometer
Electron Lett. **10** 425

Veen R J 1979 Piezojunction effect of a planar n–p–n transistor for transducer aims
Electron. Lett. **15** 333–4

Verster T C 1972 The silicon transistor as a temperature sensor
Temperature, its measurement and control in science and industry **4** (Pittsburgh: Instrument Society of America) 1125–34

Wlodarski W 1973 Field-effect transistor as a hydrostatic pressure transducer
Bull. Ac. Polon. Sc. **21** 117–20

Yeh K W and Muller R S 1976 Piezoelectric DMOS strain transducers
Appl. Phys. Lett. **29** 521–2

Zieren V 1980 A new silicon micro-transducer for the measurement of the magnitude and direction of a magnetic-field vector
Proc. IEDM Washington 669–72

Chapter 5

Digital transducers

G A Woolvet

School of Mechanical, Aeronautical and Production
Engineering, Kingston Polytechnic, Kingston-upon-Thames,
Surrey KT2 6LA, UK

Abstract. This article introduces some of the techniques used in
transducers which are particularly adaptable for use in digital
systems. The uses of encoder discs for absolute and incremental
position measurement and to provide measurement of angular
speed are outlined. The application of linear gratings for
measurement of translational displacement is compared with the
use of Moiré fringe techniques used for similar purposes.

Synchro devices are briefly explained and the various
techniques used to produce a digital output from synchro
resolvers are described.

The article continues with brief descriptions of devices which
develop a digital output from the natural frequency of vibration
of some part of the transducer. The final section deals with
descriptions of a range of other digital techniques including
vortex flowmeters and instruments using laser beams.

1. Introduction

The increasing use of digital systems for measurement, control
and data handling leads naturally to a need for transducers
which provide a digital output. A digital output from a
transducer enables direct acquisition of the output by a digital
system and simplifies processing for a digital readout or for
control purposes.

Unfortunately, nature has not provided any phenomena that
give a reasonably detectable output in directly digital form. The
only possible exceptions to this are those devices in which the
frequency of free vibration varies in response to some change in
a physical characteristic experienced by the device.

Most transducers used in digital systems are primarily
analogue in nature and incorporate some form of conversion to
provide the digital output. Many special techniques have been
developed to avoid the necessity to use a conventional analogue-
to-digital conversion technique to produce the digital signal. This
article describes some of the direct methods which are in current
use of producing digital outputs from transducers.

1.1. Transducers in digital systems

Systems based on a central processor and using a number of
transducers can use the processor as an intelligent interrogating
systems and can, therefore, make use of conventional
transducers each providing a DC output. For example, the
system can be programmed to identify the transducers in
sequence and use a single analogue-to-digital converter to
provide a digital signal representing the transducer output. The
data can be stored for further use or programmed to provide a
digital readout.

Such systems are becoming progressively more economical
with the relatively decreasing cost of microprocessors and
integrated interface devices.

An alternative approach is to use compact electronic
packages that can be housed within the transducer. The package
can provide all the sampling and control circuits for the primary
analogue signal developed by the transducer element and
convert this directly to a digital output. Such instruments will
have the appearance of digital transducers. The overall accuracy
and resolution of the instruments will now be associated with the
characteristics of the transducer element itself and with the
characteristics of the digitising electronics.

Use of the digital output signal from any instrument
containing its own analogue-to-digital conversion normally
presents no difficulties if a digital readout only is required.
However, to access the digital output by another digital system,
such as a microprocessor, will generally require some
'handshake' control. This will be necessary to prevent updating
of the digital information from the A/D conveter during the time
that transfer of information between the systems is actually
taking place.

45

Much greater flexibility can be introduced by using a dedicated microprocessor-based system in each instrument. This method has its particular uses where the instrument may be working remotely from the central control or data acquisition system but where a local independent digital readout is also required.

For systems involving a large number of reasonably local transducers it is more economical, at the present time, to have all the conversion and control organised centrally by a master processor. The transducers can then be conventional analogue types and the central processor used for some intelligent assessment of the incoming signals. The computer could be programmed to make allowance for nonlinearities and for signal recovery where the connection from the transducers have given rise to noisy signals.

Although microprocessor systems can accomplish a great deal, the obvious and most direct method in any system involving digital control, or digital readout, is to have transducers that develop an output which is directly available in some binary coded form or requires the minimum of additional electronics to provide such an output. Devices which meet this criteria are generally referred to as digital transducers and some of these are described below.

2. Angular digital encoders

Shaft encoder discs of the type shown in figure 1 were originally developed for direct electrical contact. The dark areas represent

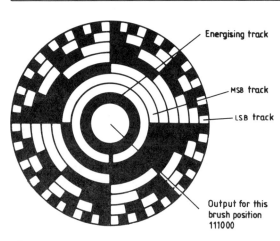

Figure 1. An absolute encoder disc.

conducting material through which a brush could be made to complete an electrical circuit. A separate brush is required for each of the concentric rings or tracks, each track representing a separate 'bit' of the digital output with an additional brush as the 'common' connection on the energising track.

Most of the shaft encoders in use today, however, make use of optical and photoelectric devices. The discs are basically transparent with opaque areas arranged in concentric rings. A common light source on one side of the disc illuminates a stack of photocells on the other side, usually with one photocell for each ring. Some optical elements are always included so that the tracks are viewed by the system through a narrow radial slit (figure 2). The output of the cells will be a parallel binary coded signal which represents the absolute shaft position according to the opaque or transparent areas of the rings along the radial slot.

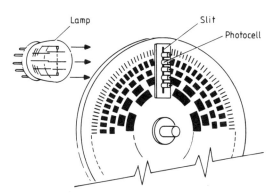

Figure 2. Optical shaft encoder.

In the normal absolute encoder the resolution is equal to the number of tracks since each produces one 'bit'. To obtain a resolution of small angles of arc, therefore, requires a large number of tracks which results in a relatively large diameter transducer. The electrical output of each cell must be conditioned and converted to a square pulse for positive interrogation purposes.

In a natural binary output each track represents one 'bit' of the output, with the inner track representing the most significant 'bit' and the outer track the least significant 'bit'. This has the disadvantage in that at some positions many 'bits' are required to change simultaneously for a small angular displacement of the disc. For example, changing from 00111111 to 01000000, representing a minimal detectable angular change, requires seven of the eight 'bits' to change simultaneously. Since it is physically impossible to manufacture an encoder disc or assemble its optical system such that all 'bits' change simultaneously then any reading taken during a change-over could be false, This problem can be overcome by using disc having a cyclic code, e.g. Gray (Woolvet 1977) code or for a continuously rotating shaft, a memory which stores the reading and changes only at the midpoint of each least significant 'bit' of the outer track.

Absolute encoders of the type described above are available with resolutions up to 14 bits representing approximately 0.02 of a degree of circular arc or 1.3 minutes of arc. Changes in position can be detected at rates up to 0.5 Megawords per second. For rotating discs, the maximum permissible speed for accurate position measurement depends upon the resolution and the switching speed of the associated electronics. Generally the greater the resolution the lower is the maximum rotational speed that can ensure an accurate read-out. For example, a 14 bit resolution with electronics capable of detecting 0.5 Megawords per second has a limiting speed of about 1500 RPM before changes in position occur faster than can be detected. Only when the shaft speed is less than this can an absolute encoder of this type output the true absolute value of the shaft position.

2.1. Optical resolver

By suitable modification to the optics and/or the electronics, the output of the cells of the outer track of an angular digital encoder can be made to produce a sine wave output. By suitable resolving, this output can be made to yield additional resolution. In practice an additional track is usually added radially outside the outer track of the absolute encoder and used exclusively as part of the resolver system. This additional track would normally have twice the number of bits as the outer track of the absolute encoder.

One sensor of this additional track is positioned such that the electrical output will vary sinusoidally in magnitude as the disc is displaced by one complete 'opaque-transparent-opaque' cycle. A sinusoidal output can be achieved by careful design of the optics of the detector, particularly, for example the shape of the aperture through which the sensor is illuminated. A second sensor is positioned such that its electrical output is displaced electrically by 90° from the first sensor. It represents a cosine output relative to the sine output of the first sensor. In practice

pairs of sensors are often used to produce each of the two signals.

The sine and cosine outputs can be processed by an interpolator to provide up to 16 or more intermediate sine waves, that is 16 sine waves differing in phase equally over the 360° electrical degrees represented by one digit of the least significant bit of the absolute encoder. Each separate sine wave can be processed to provide two output pulses.

Sixteen interpolated sine waves can, therefore, provide 32 additional divisions of the least significant bit. This is equivalent to an additional resolution of 5 bits. Figure 3 illustrates the arrangement of a transducer of this type. This has an encoder disc with 14 inner tracks providing a 14 bit absolute output. The total resolution is increased by the interpolator to 19 bits representing a shaft displacement of less than 2 second of arc. The angular displacement represented by the last 5 bits cannot provide a truly absolute output. A counting system is necessary to count the number of the 32 generated divisions displaced by the rotating disc from the last change in the least significant bit of the 14 inner tracks of the absolute encoder. However, once the system is operating, i.e. the shaft has rotated, at least by the amount represented by the least significant bit of the 14 track encoder, then the total system should always output a total absolute value of the shaft position whether it is stationary or rotating. Some further details on interpolating systems are given in § 3.3.

2.2. Incremental shaft encoders

An incremental encoder disc, uses a single track, usually with optical detectors to provide sine and cosine outputs, similar to the resolver output described above. In addition, a single mark on the disc with an associated optical pick-up is used as an

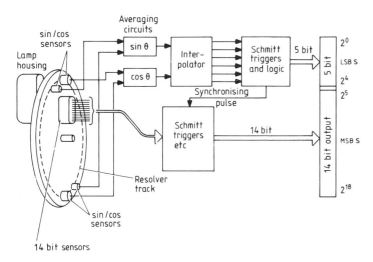

Figure 3. 19-bit shaft encoder.

angular datum from which the actual angular position of the shaft can be determined. One of the outputs of the sine/cosine pair is used to count the number of bits to or from the datum. Together the sine/cosine pair are used to determine the direction of rotation (Woolvet 1977) and hence an up or down count of the counter.

The current counter output represents the absolute value of the angular position of the shaft from the datum. When starting, the encoder disc must pass the datum position before the digital output has any validity. The two separate outputs, sine/cosine, also provide opportunity for increasing the resolution, either by interpolation as described above or by similar techniques. The simplest method is to use the leading and trailing edges of both outputs to develop a count pulse, thus increasing the resolution by a factor of four. A disc with 5000 segments, which is about the maximum currently available, on a disc 150 mm in diameter, can, therefore, provide 20 000 pulses in one revolution without any resolver techniques, which represents a resolution of approximately 1 minute of arc. In incremental encoders as in any other position encoder using counting and/or triggering circuits, there are limitations on the speed of rotation due to the maximum switching speeds of the electronics. However, for rotational speeds found in industrial situations, suitable electronics is available.

2.3. Digital tachometers

Any device which can generate an electrical pulse or a series of pulses for each revolution of a shaft can be used to initiate a digital output of the shaft's speed of rotation. The simplest device is one using a photocell to detect the passing of a white or bright mark on the shaft from reflected light. A toothed wheel using an electromagnetic pick-up, or a capacity sensitive network and a simple probe, can also provide the necessary pulses. Greater accuracy can often be obtained by using an incremental optical encoder disc with a single photocell detector. Using two detectors and the techniques employed in shaft position encoders the direction of rotation can also be determined. In all cases a counter/timer circuit is required and two basic techniques are currently used. The selection of the method to be used is governed by the speed range and accuracy required.

The first method uses a clock as a timer to count the pulses developed by the tachometer over a given period of time. This can only provide the average speed over the measurement time period. Further, the digital output available represents the average speed over the previous time period since the counter must be used to count the pulses during the current time period. The accuracy of this method depends upon the accuracy of the clock providing the time period and the length of the time period chosen relative to the actual shaft speed. The resolution of the output increases linearly with the shaft speed and therefore very low speed measurements are not really possible.

The second method uses the time between successive tachometer output pulses as the time period over which a high

clock rate is counted. In this method the resolution is inversely proportional to the shaft speed and high speed measurements are often not possible. It has the advantage, however, that it gives the average speed between each tachometer output pulse and if there are a number of pulses per revolution the angular speed at various shaft positions can be determined.

A third method, not often used in practice, is to make use of the output of an absolute encoder. For this method it is necessary to sense the time interval between two or more shaft positions. The direction of rotation is also required. A separate counter is necessary to accumulate clock pulses at a known rate between the two shaft positions. This enables the average shaft speed, between these positions, to be determined.

3. Linear displacement transducers

The digital measurement of linear (or translational) displacement can be achieved by mechanical conversion of the linear motion to rotary motion and then using a rotary digital transducer such as a rotary encoder. Various methods are currently used but precautions must be taken to avoid errors arising from backlash and nonlinearities which will reduce the overall accuracy compared to the digital transducer itself. Many computer controlled machine tools use ball-screw drives to move the horizontal slides with rotary encoders on the lead screws. These devices are extremely accurate and backlash problems have been reduced to levels representing less than 0.002 mm of linear travel

3.1. Optical gratings

A more positive approach for measurement of linear motion is to use a direct linear encoder track or scale providing either absolute or incremental output. The most popular types in use are those using optical techniques many with tracks of opaque and transparent areas. Basically the 'scale' is a straightened version of a rotary shaft encoder. Whilst it is possible to have multitrack absolute encoders the 'scales' become very wide and difficult to install and maintain. Most 'scales' are of the incremental type and consist of finely divided optical gratings. Manufacturing techniques have been developed which enables scales up to 10 m in length to be made with a grating pitch of 0.01 mm with a very high degree of accuracy.

Two measurement systems are in current use, in both cases, the scales move relative to the optical system. One uses transparent scales illuminated one side with light sensitive cells on the other as shown in figure 4. The other method uses a reflective technique with the illumination and the cells on the same side and the scales usually polished steel with the grating engraved or etched on the surface. The optical system is arranged such that it is the reflected light from the scale which is detected by the cells. Simply counting pulses generated by the grating scale leads only to a coarse resolution. The systems actually in use, therefore, usually involve some form of resolver/interpolation to increase the resolution similar to that described above for the subdivision of the outer track of a shaft encoder. This requires multiple outputs electrically phased to provide at least sine and cosine outputs. The phasing is relative to the sinusoidal output derived from one photocell output. This type of output, (sine/cosine), is also necessary in order to derive the direction of motion. In addition, there must be some form of extra marking on the scale to act as a position datum. This datum usually is at one end of the scale and is the point where the counter would normally be set to read zero. In some systems the counter can be set to zero at any give position along the scale. This position then becomes the datum from which all measurement is determined.

In most linear transducers the optical system is arranged to

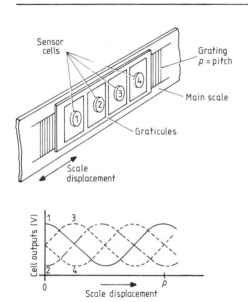

Figure 4. Linear grating assembly.

have four simultaneous separate outputs. In one the output will vary sinusoidally by one cycle as the optical system moves one 'grating' distance. A second output is arranged to produce the inverse of the sine, i.e. '−sine'. The third output is so positioned to produce the 'cosine' output, that is displaced electrically by 90° from the sine output. The fourth provides a negative cosine, i.e. a '−cosine' output.

Each of the four outputs is obtained by detecting the light passing through the scale and a graticule. The graticule consists of a short length of transparent grating having the same pitch as the main scale (figure 4). The photocell will have maximum output when the ruled lines of the scale and graticule coincide and minimum output when the lines of the scale coincide with the spaces of the graticule. The photocell behind the graticule averages the light received over a short length of the scale, i.e. over a number of grating lines of the main scale. This further

minimises any small errors that might exist in the spacing of the grating. Four separate graticules are used on a single indexing frame, positioned relative to each other, to provide the four phased outputs as the graticule and optical assembly moves relatively to the scale as shown in figure 4. These outputs are used in interpolator networks to provide four or more subdivisions of the main scale division.

3.2. Moiré fringe techniques

Moiré fringes are produced by a transparent grating and graticule, both having the same pitch, but with lines of the graticule inclined at a very small angle to the lines on the scale. Figure 5(a) shows the effect produced when the graticule is inclined by one grating pitch, (p) over the width of the scale. The angle α is given by;

$$\alpha = \tan^{-1}(p/y).$$

In this case there is one horizontal dark area which moves vertically the distance y, for a horizontal displacement of the scale by a distance p.

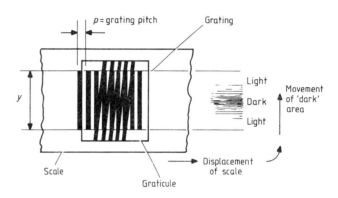

Figure 5. Moiré fringe.

Two horizontal dark areas are produced if the angular incline of the graticule is $2p$. Similarly, three horizontal dark areas are produced by an inclination of the graticule of $3p$ and four horizontal dark areas by an inclination of $4p$.

A single light sensitive cell at a fixed vertical position across the width of the main grating will sense one complete 'dark–light–dark' cycle as the graticule is displaced by a distance p.

By careful positioning of two separate light cells the output can be made to represent sine and cosine of one cycle of displacement, that is, the two outputs are separated by 90° electrical degrees. These two outputs can then be used to increase the resolution by a factor of four in a similar manner to the methods used in incremental shaft encoders. Operating on a grating with 100 lines for each linear millimetre, this will provide a resolution of 0.0025 mm. The direction of movement can also be determined from these two outputs, in order to direct the counter to count up or down.

Four light cells can be positioned to provide the sine/cosine outputs and their negative values as described in § 2 in relation to shaft encoders. Interpolation of these four outputs can use the techniques used for many other incremental techniques where similar output can be produced.

3.3. Interpolation systems

The simplest interpolation network consists of a resistor chain as shown in figure 6 to which the sine/cosine outputs are connected. The outputs A, B, C, D, & E, produce a set of quasi sinewaves, differing in phase according to the tapping position on the resistance chain. Each separate waveform from the pattern is then used to generate a square wave, by the use of level detectors and these differ in phase relationship as shown in figure 7. By using a relatively simple logic system, ten separate pulses can be generated. This method provides a tenfold increase in resolution over the scale grating.

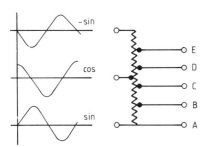

Figure 6. Simple interpolation network.

4. Synchro-resolver conversion

Synchros are used extensively in military control systems and are finding increasing use in numerically controlled machine tools and other industrial applications. Although most of the various synchro devices are more suited to AC control systems, a number of different conversion techniques have been developed to provide a digital output of a shaft position.

Resolvers are effectively rotary transformers in which a rotor represents the primary and is fed through slip rings by a single phase reference supply. The stator, which is in effect the secondary, contains two windings which are arranged to produce two separate voltages, at the same frequency as the reference supply. The magnitude of the voltages on the two stator windings varies as the angle of displacement of the rotor relative to a datum on the stator. The magnitude of one voltage varies as the sine of the angle of the shaft position and the other

as the cosine. The two outputs are, therefore:

$$v_1 = V_m \sin \omega t \sin \theta \qquad (1)$$

$$v_2 = V_m \sin \omega t \cos \theta \qquad (2)$$

where

$$\omega = \text{frequency of reference supply}$$

$$V_m = \text{maximum output voltage}$$

$$\theta = \text{angular position of rotor.}$$

The supply or reference frequency may be 50 Hz but is more usually 400 Hz in instrumentation systems and higher for special purposes.

The two resolver outputs are in effect amplitude modulated

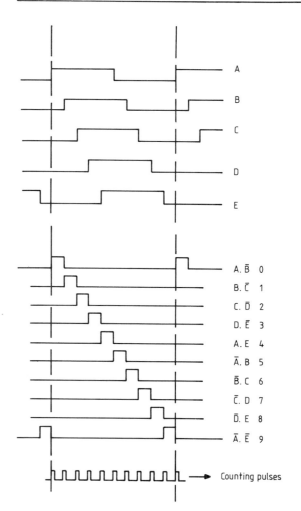

Figure 7. Interpolator logic.

signals at the reference frequency. These signals are demodulated or 'converted' by one of the following methods:
(i) phase-shift converters
(ii) function-generator converters
(iii) tracking converters
(iv) successive approximation converters.

The aims of each are similar, that is to provide a digital output proportional to the rotor position. The information contained in the two resolver signals is sufficient to define uniquely the position of the rotor relative to the stator over the full 360° of rotation. The various conversion techniques all use the two analogue signals to produce a digital output. It is the converter that makes the resolver into an incremental transducer producing pulses which have to be counted to provide digital

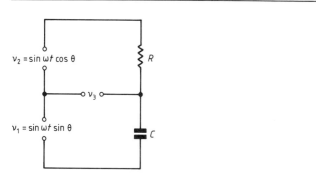

Figure 8. Phase-shift converter. $\omega RC = 1$, where ω is the reference frequency.

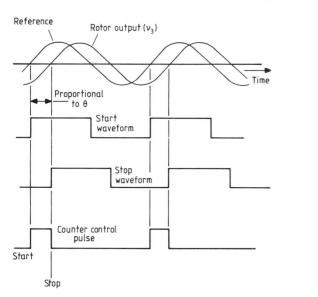

Figure 9. Phase-shift converter waveforms.

position data. The differences between the various converter methods is in the resolution available, the speed at which the shaft can be rotated and still maintain the designed resolution and the sensitivity of the system to the unwanted distortion of the resolver signals.

4.1. Phase-shift converters
If the two resolver output equations (1) and (2) are fed into the circuit shown in figure 8 the voltage v_3 will be given by:

$$v_3 = \sin(\omega t + \theta).$$

By changing this sinusoidal voltage and the reference voltage to square waves, the two waveforms can provide a start and stop signal for a counter system as shown in figure 9. The number accumulated in the counter then represents the phase shift of v_3 relative to the reference supply and therefore the angle of rotation of the rotor, θ. It is important that the reference frequency and the count rate are synchronised since the real time between the start and stop pulses is a function of the reference frequency. The resolution depends upon the ratio of count rate to reference frequency. However, to maintain this resolution the speed of rotation of the shaft is limited since the counter update takes place only at the reference frequency.

Generally, therefore, this method is used only for slow rotational speeds; 20 rev/min is a typical maximum speed for a resolution of about 1 degree of arc. Special techniques have been developed to overcome the synchronisation problems and to increase the resolution and the operating speeds but for significant improvement other conversion techniques are usually adopted.

4.2. Function-generator converters
These converters use a feedback technique in which the digital output is fed back to a function generator and a comparator which develops a signal proportional to the error between the shaft position and the digital output. The error signal is used to drive the digital output towards a value which reduces the error to zero (figure 10).

Even in its simplest form this system provides a higher degree of accuracy and resolution (± 5 seconds of arc) although more complex and more expensive than phase-shift converters. It is also less sensitive to unwanted components in the resolver signals and to variations in the reference supply.

There are a number of methods of achieving the 'function generation', but most are effectively hybrid multipliers which generate an analogue output signal which is the product of analogue inputs, (representing $\sin \theta$ and $\cos \theta$) and a function of a digital input representing the encoder output φ.

The outputs of the function generators are therefore:

$$\nu_a = \sin \omega t \sin \theta \cos \varphi$$

$$\nu_b = \sin \omega t \cos \theta \sin \varphi.$$

51

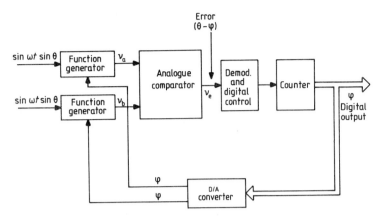

Figure 10. Function generator converter.

These two voltages are fed to a comparator circuit producing an error voltage ν_e proportional to $(\theta - \varphi)$.

$$\nu_e = \nu_a - \nu_b = \sin \omega t (\sin \theta \cos \varphi - \cos \theta \sin \varphi)$$

$$\therefore \nu_e = \sin \omega t \sin(\theta - \varphi).$$

The voltage ν_e is demodulated to produce an analogue signal which is a function of $(\theta - \varphi)$ only and is used to control a counter to produce the digital output. When $\theta = \varphi$, the error $(\theta - \varphi)$, and therefore ν_e, becomes zero and the counter value remains constant and the output is then the digital equivalent of the encoder shaft position θ.

The simplest form of function generator uses a linear resistor network similar to that shown in figure 11. The output voltage achieves the required function by the digital input selectively switching the voltage divider network.

4.3. Tracking converters

These are function generator converters which adopt a particular technique for generating the digital output. The demodulated error signal in analogue form is integrated with respect to time and the result used to drive a voltage to frequency (v/f) converter. The resulting serial output is fed to a counter whose output represents the value of φ. This digital signal is fed back to the function generators as previously described. When $\theta = \varphi$ the error voltage is zero, the frequency output also zero and, therefore, the counter and hence the digital output remains constant. A tracking converter produces a digital output which remains equal to the rotor position whilst the shaft is stationary or rotating at constant velocity.

The block diagram, figure 12 shows the counter as an integrator and indicates the feedback aspects of the system. The double integral in the forward path formed by the analogue integration of the error signal and the counter, defines the steady state and transient characteristics of the complete converter loop. These are:

(i) no steady state error, i.e. $\theta = \varphi$ when θ is unchanging.

(ii) if the input θ is changing, (i.e. the rotor shaft turning) φ will also be identical with θ and there is no errors due to velocity of θ.

(iii) if the input θ is accelerating the digital output φ will lag behind the input θ.

Typical tracking encoders can maintain a 14 bit resolution (1.3 minutes of arc) up to speeds of 240 rev min^{-1}. Higher speeds will limit the resolution and *vice versa*. If the encoders are given a step input, e.g. the input–output error is 180° of shaft rotation, the time taken for the output to represent the input depends upon the maximum frequency of the voltage-to-frequency converter and the time taken could be up to 0.5 seconds. Special techniques are used for increasing the resolution and adding other refinements to the overall performance.

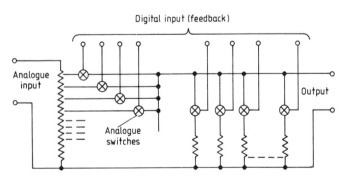

Figure 11. Function generator.

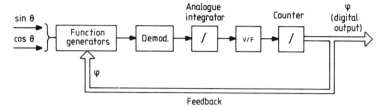

Figure 12. Tracking converter block diagram.

4.4. Successive approximation converters

These are very similar to the tracking converters discussed above except that the output of the voltage-to-frequency converter is converted to a digital output by a successive approximation method similar to those used in analogue-to-digital converters. This system can be faster than the tracking converters.

4.5. Harmonic oscillator converters

The central feature of this type of converter is an harmonic oscillator which is an analogue circuit (figure 13) with two integrators and if left free running would oscillate at a frequency in the range of 100 to 200 Hz. The outputs of the integrators represent the sine and cosine of the oscillator waveform. The two resolver outputs, proportional to $\sin \theta$ and $\cos \theta$, are demodulated and the resulting two DC outputs, whose magnitudes represent the $\sin \theta$ and $\cos \theta$ respectively, are used to set the initial conditions of the two integrators of the harmonic oscillator.

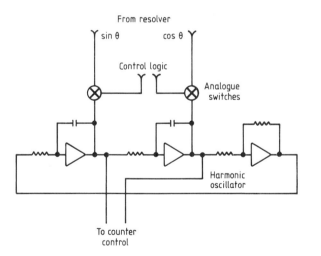

Figure 13. Harmonic oscillator converter.

The converter includes FET switches to switch these initial conditions to the integrators, a clock, a counter and control logic. The complete measurement cycle starts by stopping the harmonic oscillator and switching the analogue values of $\sin \theta$ and $\cos \theta$ to set the initial conditions of the two integrators. The programming logic then sets the harmonic oscillator running at its natural frequency with the analogue outputs at the initial conditions previously set, simultaneously isolating the $\sin \theta$ and $\cos \theta$ inputs. At the same time the counter is initiated and clock pulses are counted until either of the harmonic oscillator outputs cross zero. This has the effect of providing a count equal to the initial value set on the integrators and therefore equal to the magnitude of θ, the resolver shaft angle.

The counter information is up-dated periodically at the harmonic oscillator frequency e.g. approximately 200 conversions per second. With a clock rate of 1 MHz the output could have a total resolution of 12 bits including those bits which are derived from addition logic which identifies the quadrant (Woolvet 1977) in which θ exists. Overall accuracy depends upon accurate clock rate, stability of the harmonic oscillator and a number of other characteristics which can usually be improved by further sophistication of the system.

Synchro-encoders of any type are effectively absolute devices always giving the angular position of the rotor over 360°, relative to an electrical datum of the stator. If the encoder is used for continuous rotation then an additional counter system must be used to take account of complete revolutions, and the $\sin \theta$ and $\cos \theta$ outputs used to determine the direction of rotation.

5. Variable frequency devices

There are a number of techniques in which the variation of a parameter to be measured can be used to cause the variation in the natural frequency of vibration of some part of the transducer. The actual measuring method will be similar in many ways to simple incremental encoders in that a counter/timing circuit is required. Either the periodic time of the vibration frequency may be measured by counting pulses from a high frequency clock or alternatively the frequency output can be converted to pulses which can be counted over a given time period.

The common feature of the type of transducer described in this section is that the change in frequency does not depend upon relative motion between parts of the transducer as in encoder systems. Almost any parameter which can be measured by a change in DC potential or change in resistance can be used by a voltage-to-frequency converter, or be formed as part of an oscillator circuit, to provide a variable frequency output which can then be used to develop a digital output. Examples of these systems are, a strain gauge bridge on a diaphragm for measuring pressure, potentiometers for measuring displacement and thermistors for measuring temperature.

5.1. Vibrating strings and beams

Vibrating string transducers have taken a variety of forms, the most popular being used as strain gauges and have been used to measure both force and strain. The string or wire, usually steel, is fixed at one end and the force to be measured applied at the free end. As a strain gauge both ends would be fixed to the structural member whose strain is to be measured. Any change in tension in the wire caused by a change in the force applied will change the natural frequency of free vibration. The frequency change is measured by a variable reluctance pick-up which is amplified to provide the frequency output and also to provide a feedback to an electromagnetic exciter to sustain the wire in free oscillation as shown in figure 14.

The frequency of free vibration of a taut wire or string is given by:

$$f = \frac{1}{2L}\left(\frac{T}{m}\right)^{-1/2} \text{Hz}$$

where

L = length of wire between supports
T = tensile load
m = mass per unit length.

A thin beam, that is, a small steel tape, can be used in a similar manner. An alternative application has used a twisted beam or tape stretched across the faces of the electromagnetic pick-up and exciter. The frequency changes as the beam is twisted and the transducer therefore acts as the measurement of angular displacement.

5.2. Vibrating cylinders

However, the most popular of transducers under this heading are those incorporating vibrating cylinders which have been commercially developed for the measurement of fluid pressure, density and mass flow. In the most common design the measuring element consists of a steel or alloy cylinder closed at one end approximately 25 mm diameter, 50 mm long with the side wall only 0.075 mm thick. The natural frequency of vibration selected is usually in one of its most stable modes,

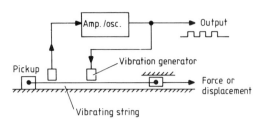

Figure 14. Vibrating string transducer.

which is sustained and detected by iron cored solenoids across the diameters of the cylinder. An amplifier and filter system and the position of the exciter ensures that the cylinder vibrates only in the mode selected.

The frequency of vibration depends upon the dimensions and the material of the cylinder and upon any mass caused to vibrate with the cylinder walls. Since the gas in immediate contact with the walls of the cylinder also vibrates the frequency depends upon the density of the gas. In this case the transducer will output a frequency and hence a digital output which is a function of the gas density. In this application the gas pressure must be the same on both the outside and the inside of the cylinder. Any differential pressure across the cylinder wall will create a tension in the cylinder and change the frequency of free vibration. This factor allows the cylinder to be used as a pressure transducer providing one side of the cylinder is maintained at constant pressure, (e.g. a vacuum).

A typical pressure transducer measuring up to 300 atmospheres will have a natural frequency varying from 1.5 kHz to 5 kHz over its full working range. The cylinders will be of different physical size for different ranges. Special precautions must be taken to minimise errors due to temperature change and also nonlinearities in the pressure and the frequency output.

Another type of vibrating tube transducer, larger in physical size than the pressure transducer, is used for measuring liquid density. The liquid is caused to flow through two parallel pipes whose ends are secured together and to a rigid base plate. The tubes are coupled by flexible couplings to the main flow. Between the tubes are the drive and pick-up electromagnetic coils which cause the tubes to vibrate in a simple lateral mode. The frequency of free vibration is a function of the density of the liquid filling the tubes.

Other applications of vibrating cylinders have been suggested including the measurement of force and the measurement of torque.

6. Other techniques

6.1. Depth transducers

A direct digital output of fluid depth can be obtained either by using transducers to detect the change of inductance or change of capacitance. For the inductance types, the transducer elements consist of pairs of coils located either side of a vertical tube. The coils are located outside the tube. The inductance across the tube between a pair of coils is influenced by the presence or absence of liquid in the tube separating the coils. Nearly all fluids will cause sufficient change of inductance to produce a signal change. A number of such pairs of coils may be located along the length of the tube each pair designed to provide a trigger when the fluid reaches a level adjacent to the coils. The resolution is only equal to the number of such pairs of transducers. Also, the system can only sense large increments in depth due to the minimum vertical separation of the coils necessary by their physical size. However, since the system is in effect a digital manometer the sensitivity can be greatly increased by inclining the tube towards the horizontal. This leads to a reduced detectable range. The method could also be used as a pressure gauge by using a capsule whose displacement caused by a change of pressure is used to displace the fluid along the tube. Similar limitations apply to the capacitance sensors since they are used in a similar way. In place of pairs of coils, there would be pairs of plates representing a capacitor, the actual capacity depending upon the dielectric. In some cases the capacitors may be completely immersed in the fluid whose depth is to be measured. Both inductive and capacitive systems require an AC modulation and demodulation system which can add considerably to the overall cost.

6.2. Magnetic effects

The magnetic recording of computer data, on tapes and discs, has been developed to densities up to 200 bits per mm and greater densities are theoretically possible. Since it is possible to record eight or more tracks simultaneously it would seem that there are considerable possibilities for direct digital readout of any parameter which can be measured as a displacement.

Two magnetic systems are currently available both used for the measurement of translational displacements and both working in a similar way. One method employs a flexible metal tape and the other an alloy rod. The rod or tape is prerecorded, with a continuous track of bits, which is moved relative to fixed 'replay' head or heads. The replay heads detect the passing of the recorded bits which provide an output which can be summed incrementally. Two or more heads may be used to provide the sine/cosine outputs for interpolation to improve the resolution (3.3). The maximum displacement velocity is limited by the quality of the recording on the tape or rod. For example a tape or rod velocity of 50 mm s^{-1} (corresponding approximately to the standard domestic cassette tape speed of 1.875 inches per second) and prerecorded to a density of 200 'bits' per mm is equivalent to a recording frequency of 10 kHz. Higher velocities would therefore require good quality recording on the tape or rod and very high precision 'reading' heads.

6.3 Radiation transducers

The random radiation from a radioactive source can be detected by a photomultiplier or other type of detector. The random series of electrical pulses generated can be counted over a given period and provide a digital output proportional to the radiation received by the detector. A counting period in the order of 1 s has been found in some practical cases, to give a repeatable resolution of 0.1%. The radiation received by the detector depends upon the strength of the source, the distance of the source and the area of exposure. The long life of a properly selected source and a fixed distance between source and detector reduces the detector output as a function of the area of exposure only and this can very conveniently be made to vary as a function of linear displacement or angular rotation. Practical transducers working on this principle are very rugged and more independent of many environmental effects than many other types of transducer.

In some installations it is possible to use radiation techniques for measurement of fluid flow. A neutron source located upstream of a fluid flow will cause some of the water particles to become radioactive and some of these can be sensed by a detector further down stream. For a given fluid, pipe size and given distance between the source and detector, the detector count gives a direct measurement of the flow rate. As with any counting technique, resolution and accuracy is dependent on the time period of the count.

6.4. Vortex transducers

There have been a number of techniques used for determining flow by the natural vortices caused by interference in the smooth flow of a fluid. The method now successfully developed commercially measures the flow rate in pipes by measurements of the vortices developed by a thick strut placed across the diameter of the pipe. The flow is disturbed and above a minimal velocity a regular series of vortices will be shed, the frequency depending upon the velocity.

The sensing element is required to detect the frequency of the vortices as they pass a detector located immediately down stream of the strut. The most successful method uses an ultrasonic transmitter radiating across the flow and a detector which senses when a vortex passes, by a change in the level of the signal received. This change can be used as a trigger to start and stop a counting circuit to provide a digital output measuring the time between successive vortices and therefore the fluid flow rate.

Other means of detecting the vortices such as sensitive piezoelectric pressure transducers to detect the pressure change in a vortex have also been used.

6.5. Laser techniques

The collimated beam produced by a continuous wave laser can be used in many instrument systems, particularly for measurement of the relative displacement of large structures and for measurement over long distances. For a digital output of measurement of the shorter distances commonly associated with metrology and instrumentation generally requires the use of a laser interferometry system. The fringes produced by a laser interferometer are similar to the Moiré fringes referred to earlier but in the laser system they are due to interference of the laser beam and a reflected beam made to follow the same path as the original beam. The effect is to produce a dark fringe when the two coincident waves are out of phase and a light fringe when in phase. Any movement of the reflecting surface which changes the distance travelled by the reflected wave will produce a fringe. The fringe pattern is focused on to photocells in a similar manner to that used with Moiré fringes. One complete cycle of interference corresponds to a movement of half a wavelength and this would be in the order of 0.001 mm.

For measurement purposes a counter would be required to count the number of fringe changes. A second sensing element would also be required to produce a quadrature signal in order to determine the direction of movement in order to generate a signal to initiate a count-up or count-down. Some compensation may also be required to overcome the variations caused by changes in temperature and air pressure. Instruments of this type are expensive but have been used to measure minute changes in displacements over long distances and also for very small displacements caused by adhesive films or thermal expansions.

A second method uses two laser beams and is, therefore, even more expensive. The two lasers must emit beams of slightly different wavelength and in opposite polarisations. The two beams are reflected from the surface whose displacement is to be measured and compared in an interferometry circuit with the beams directly from the lasers. The beat signal of the reflected laser beams will vary with the motion of the reflecting surface and the difference between the beat frequency of the reflected beams and the beat frequency of the direct beams is a measure of the velocity of the reflecting surface. The difference is a Doppler shift and if a counter is used to count the difference in frequencies a measure of the reflector displacement is obtained.

7. Conclusion

The principles described above cover the majority of the techniques which have been adapted for use in digital techniques. Whilst it is difficult to suggest future developments it is safe to forecast that other devices will be developed. Most of these will undoubtedly be adaption or combinations of methods already existing. The most valuable advance will no doubt be made by the invention of relatively simple devices to provide the digital outputs required.

Acknowledgment

The illustrations in this paper are reproduced courtesy of Peter Peregrinus Ltd.

Reference

Woolvet G A 1977 *Transducers in Digital Systems* (London: Peter Peregrinus)

Chapter 6

Digital signal conditioning and conversion

A R Owens

School of Electronic Engineering, University College of North Wales, Dean St, Bangor, Gwynedd LL57 1UT

Abstract. The process of conversion of an analogue signal into a digital version is examined; in particular the errors introduced by the conversion are analysed. The salient features of the main methods of conversion are then described.

Since sampling of signals is inherent in the process of conversion, the mathematics of sampled signals is then examined and the discrete forms of the Laplace and Fourier transforms are developed. These are then used to show how signals may be processed in the digital domain.

Finally, the principles of the design of circuits for the realisation of digital signal conditioning algorithms are described.

Introduction

Digital signal conditioning and conversion is concerned with processing a signal by digital means prior to extracting information. Digital filtering and smoothing, signal averaging, correlation, Fourier transformation and spectral analysis are all examples of digital signal processing methods. In addition, several functions which were previously performed by analogue techniques can now be performed very effectively by digital methods owing to the ease with which they can be implemented with a microprocessor. These include phase-sensitive detection, RMS computation, integration. Finally, the generation of analogue test signals of specified characteristics can be performed with great precision with the use of digital techniques; examples include the generation of precision sine waves and random noise test signals.

1. Entering the digital domain

Although the present trend of electronic instrumentation is firmly set towards digital signal processing, most signals remain stubbornly in the analogue domain! There are relatively few examples of transducers which produce a direct digital signal from the physical input which is being measured (such as temperature, pressure, and so on); most transducers produce analogue voltage or current outputs, and as a result, the first task in designing a digital signal conditioning system is to decide on a method of obtaining a digital representation of the signal of interest. There are also some transducers which produce a frequency output which is related to the quantity to be measured; for example, capacitive or inductive transducers connected in an oscillator circuit, or a resistive transducer controlling a multivibrator.

Some of the factors which have to be considered when choosing a conversion method are (a) the number of bits required in the digital representation, with all the ramifications of quantising noise and dynamic range which this implies, and (b) the speed of conversion, which is linked with other system considerations such as bandwidth, sampling rate and interference rejection.

1.1. System considerations: noise and dynamic range

An analogue-to-digital converter is a device which accepts an analogue input x_a, and produces a digital number x_d (figure 1). x_d will be a parallel combination of M binary signals, or (less usually) a serial stream of M binary digits.

The range of x_d is from 0 to $2^M - 1$ in uniform steps; the analogue equivalent of 1 step is the quantising interval q. The full-scale analogue signal is thus $(2^M - 1)q$, and the smallest change in x_a which can be reliably observed is $\pm q$.

The relationship between x_a and its representation x_d is shown in figure 2. In (a) x_d is 00000000 for $0 \leqslant x_a < q$,

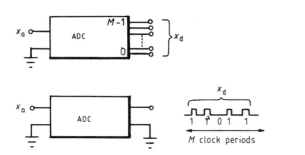

Figure 1. An ADC converts an analogue signal x_a into a digital representation x_d.

00000001 for $q \leqslant x_a < 2q$, 00000010 for $2q \leqslant x_a < 3q$ and so on. The error in the representation is $x_d - x_a = e$; in the case shown in (a) the error lies between 0 and $-q$.

In figure 2(b) the same relationship is shown with an offset of $q/2$ in x_a; thus x_d is 00000000 for $-q/2 < x_a < q/2$, 00000001 for $q/2 \leqslant x_a \leqslant 3q/2$ etc. In this case the error becomes alternately negative and positive as x_a increases, and the error statement is then $-q/2 \leqslant e \leqslant q/2$. The error is then never more than one half the quantising interval. Unless there is specific evidence to the contrary, it can normally be assumed that the error is a random quantity which is distributed uniformly over the interval $0 < e < q$ (case (a)) or $-q/2 < e < q/2$ (case (b)) (figure 3).

Using the expected value technique,

$$E\{f(x)\} = \int_{-\infty}^{\infty} f(x)p(x)\,dx \qquad (1)$$

where $p(x)$ is the probability. In this case,

$$p(e) = 1/q \text{ for } -(q/2) < e(q/2)$$

and we obtain the following results.
Mean value of e

$$\bar{e} = E\{e\} = \int_{-\infty}^{\infty} e(1/q)\,de$$

$$= -q/2 \text{ (case } (a))$$

$$= 0 \text{ (case } (b)). \qquad (2)$$

Variance of e:

$$\sigma_e^2 = E\{(e - \bar{e})^2\}$$

$$= q^2/12 \text{ (both cases).} \qquad (3)$$

This shows that even assuming a noise-free input x_a, the output signal x_d contains noise amounting to $q^2/12$.

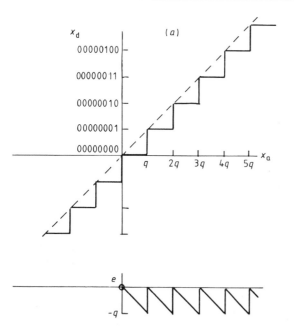

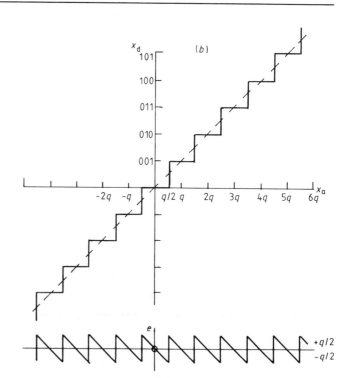

Figure 2. (a), Transfer function of a hypothetical ADC; (b), an offset of $q/2$ reduces the mean error to zero.

Figure 3. Probability distributions of quantising error.

It is instructive to compute the maximum dynamic range available. Let us assume a sinusoidal x_a biased so that the negative peak occurs at 0 and the positive peak at $(2^M - 1)q$ (i.e. a steady offset of $(2^M - 1)q/2$).

The mean square value of the sinusoid is $\frac{1}{2}((2^M - 1)q/2)^2$, and the signal-to-noise ratio of x_d is

$$S/N = \frac{1}{2}((2^M - 1)q/2)^2/(q^2/12)$$
$$= 1.5(2^M - 1)^2. \qquad (4)$$

For values of M greater than 4, the error involved in the approximation $2^M \gg 1$ is of no great consequence. Thus

$$S/N \simeq 1.5 \times 2^{2M}. \qquad (5)$$

Taking dB's,

$$S/N(dB) = 10 \lg 1.5 + 20M \lg 2$$
$$= 1.76 + 6.02M. \qquad (6)$$

This result shows that the converter produces quantising noise which is $(-1.8 - 6M)$ dB below the maximum measurable sinusoid. This defines the dynamic range of the converter for sinusoidal signals. For other signals, appropriate adjustments must be made (table 1). Note that these figures relate the noise which occurs on *any* input signal to the power of the *maximum* signal of any given type which can be accommodated. As will be shown later, digital signal averaging or filtering techniques can

be applied which can give substantial improvements in the signal-to-noise ratio.

1.2. System considerations: speed of conversion

The digital number x_d which results from a conversion of x_a is a representation of x_a at a particular instant of time: thus $\{x_d\}$ represents a series of discrete time samples of x_a. The sampling may occur periodically at constant intervals (a case which is more likely to occur if the time-varying properties or the frequency spectrum of x_a are of interest) or occasionally, at random intervals (usually where DC values, or very slowly varying values of x_a, are of interest).

In the case of periodic sampling, the constraints of the sampling process must be recognised. Basically, this means that if a broadband signal is sampled at a frequency f_s, a component at f_x will appear in the sampled output as a component $f_x - f_s$, (a process termed aliasing), which may be confused with a component at f_y (figure 4). If the component at f_y is to be analysed correctly without interference from the component at f_x, a prefilter must be employed which reduces all components above the frequency $f_s/2$ to a negligible level. Ideally the 'anti-aliasing' filter would have a characteristic

$$|H(\omega)| = 1, \ 0 < f < f_s/2$$
$$|H(\omega)| = 0, \ f_s/2 \leqslant f. \qquad (7)$$

Various compromises are achieved in practice by placing the cut-off frequency (3 dB down point) of $H(\omega)$ somewhat below $f_s/2$, and relaxing the stop-band specification.

$$\text{e.g. } |H(\omega)| < \frac{1}{2^M} \text{ for } f \geqslant f_s/2 \qquad (8)$$

(depending on the nature of the signal), the aim being to reduce the aliased components below the 'noise floor' of the converter.

Since signals can be measured without ambiguity only in the

Table 1.

Signal	Max. signal power		S/N ratio (dB)
Unipolar DC	$(2^M - 1)^2 q^2 \simeq 2^{2M} q^2$		$10.8 + 6M$
Bipolar DC	$((2^M - 1)q/2)^2 \simeq \dfrac{2^{2M}}{4} q^2$		$4.8 + 6M$
Gaussian (3σ assumed)	$((2^M - 1)q/6)^2 \simeq \dfrac{2^{2M} q^2}{36}$		$-4.8 + 6M$
Uniformly distributed signal	$((2^M - 1)q)^2/12 \simeq \dfrac{2^{2M-1} q^2}{3}$		$6M$

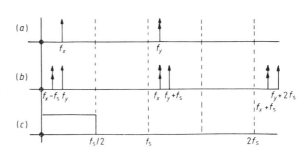

Figure 4. (*a*), Original signal spectrum; (*b*), spectrum of sampled signal (the component f_x creates a component near to f_y); (*c*), response of ideal prefilter to avoid 'aliasing' effects by eliminating f_x from the input to the ADC.

range 0–$f_s/2$ in the ideal case, or 0–$(1/k)f_s/2$ in the practical case (where k is a factor which allows a non-ideal filter to be used), this dictates the choice of sampling interval in many applications. Thus if the signal bandwidth is 0 to f_h, the sampling frequency must be $2kf_h$ or greater.

The time interval over which the input signal is sampled is also important. So far it has been assumed that

$$x_d(nT) \text{ is a representation of } x_a(t=nT).$$

In many cases, however, x_d is the result of processing the average value of x_a over a time interval τ.

$$x_d(nT) = \frac{1}{\tau} \int_{nT}^{nT+\tau} x_a(t)\,\mathrm{d}t. \tag{9}$$

The result of this is to apply an envelope $(\sin \omega\tau/2)/(\omega\tau/2)$ to the spectrum of the sampled signal (figure 5(a)).

A particular case of interest is where the signal is integrated over the whole of the sampling interval (figure 5(b)). In this case, spectral components of the x_a which lie near f_s and its harmonics are greatly attenuated, and where low-level DC measurements are involved, where interfering components at mains supply frequency may be expected, it is often an acceptable solution to apply the integration technique, using a sampling frequency related to the supply frequency, with no other anti-aliasing filters.

2. Fundamental methods of A–D conversion

2.1. Frequency measurement
The most fundamental digital method of measuring an analogue

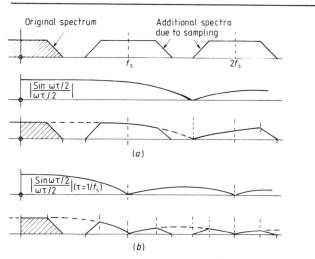

Figure 5. (a), Finite-width sampling superimposes an envelope $|\sin(\omega\tau/2)/(\omega\tau/2)|$ on the sampled spectrum; (b), when $\tau = 1/f_s$ the nulls in $|\sin(\omega\tau/2)/(\omega\tau/2)|$ tend to eliminate components near the harmonics of f_s.

quantity is the process of frequency measurement. If frequency is defined as the number of events in a given time interval, all that is required is for a counter to count the events in the stated time interval as shown in figure 6. The counter is set to zero count at the start of the operation and at the end of the measurement interval the accumulated count is transferred to another register so that the results of the last measurement can be accessed or viewed whilst a new measurement is in progress.

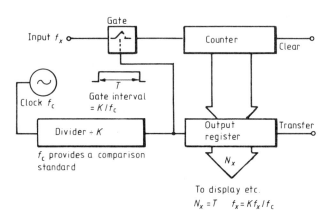

Figure 6. A typical frequency counter structure.

It should be noted that although the frequency being measured is an analogue quantity (being perhaps related to some physical effect such as the temperature variation of a capacitance sensor, for example), the counter will register the nearest whole number. Consideration of the counting process shows that the error lies between $+1$ and -1 count (figure 7).

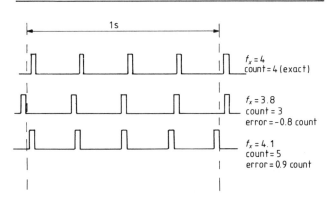

Figure 7. How errors arise in frequency counting. In all cases the best reading would be $N_x = 4$ for least error.

The variance on any one reading is therefore $\frac{1}{3}$ count. An advantage of frequency counting methods is that there is very little difficulty in cascading counters to increase the range of the measurement, at the same time extending the duration of the measurement interval. In this way, since the error range remains constant at ± 1 count, the resolution of the measurement may be made very high, but at the expense of conversion speed.

It should be noted that since the counter records the number of events in a given time interval that the reading gives the average frequency for that interval and the instrument therefore has the 'integrating' characteristics discussed in § 1.2.

An alternative way of measuring frequency is to measure its inverse, namely the period. The same components (counter, gate, clock) are used as before, but the incoming frequency is used to define the measurement interval, the signal to be counted being supplied by the internal clock (figure 8). The resolution is still governed by the maximum error of ± 1 count, but this is now fixed by the clock period.

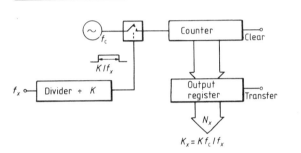

Figure 8. Rearrangement of components for period measurement.

2.2. Voltage measurement

Where the analogue quantity to be encoded is in the form of a voltage, there are three main paths which may be followed:

(i) conversion of the voltage into a frequency or time analogue, which can then be measured by pulse-counting methods, as discussed above;

(ii) a potentiometric method whereby the analogue voltage is compared with a digitally controlled voltage source which is adjusted until the best balance is attained; and

(iii) multiple comparison methods where the analogue voltage is compared (in parallel) with all possible voltage levels which can be described by the digital output codes.

2.2.1. Time analogue methods.

2.2.1.1. Single slope conversion. In a single slope converter a capacitor is charged to the unknown voltage V_x, which is then discharged by a reference current I_R to a comparison level $-V_0$. The time taken is measured by gating a clock signal into a counter (figure 9). The digital count is

$$N_x = [(V_x + V_0)/I_R]Cf_c \qquad (10)$$

and contains the unknowns V_0, C, f_c and I_R. However, by carrying out reference measurements with inputs V_R and 0, further digital counts N_R and N_0 are obtained so that the reading can be corrected to give

$$V_x/V_R = \frac{N_x - N_0}{N_R - N_0} \qquad (11)$$

thus eliminating both scale factor and zero errors. This method is suitable for use in conjunction with microprocessors, when multiplexed inputs can be scanned (e.g. Motorola MC 14447, figure 10). The conversion time varies with the amplitude of the input voltage. It is a unipolar method, unless some offset is introduced into the input.

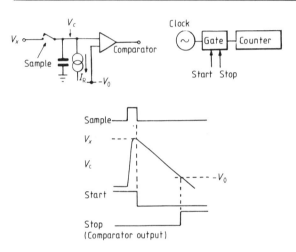

Figure 9. The single slope converter.

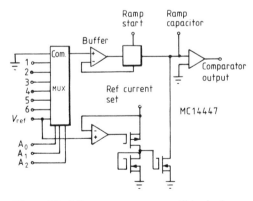

Figure 10. Microprocessor-compatible single converter integrated circuit.

2.2.1.2. Dual slope conversion. In the dual slope converter an integrating capacitor is charged up from the unknown voltage for a fixed time, determined by the system clock (phase I). During phase II the integrated circuit is discharged by a reference voltage, and the same system clock is used to measure the time taken to attain the discharged condition (figure 11).

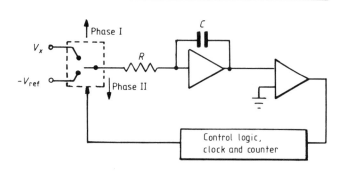

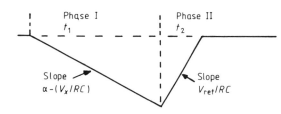

Figure 11. The dual slope converter.

Phase I is determined by the time taken to cycle the counter from 0 to N_{max}.

$$t_1 = N_{max}/f_c. \tag{12}$$

During this time the integrating capacitor acquires a voltage

$$V_c = t_1 \frac{V_x}{RC} = \frac{N_{max}}{f_c} \frac{V_x}{RC}. \tag{13}$$

During phase II the capacitor is discharged by the connection of V_{ref}, to the input, the measured time being

$$t_2 = N_x/f_c. \tag{14}$$

By equating charge gained during phase I to charge lost during phase II,

$$\frac{N_{max} V_x}{f_c RC} = \frac{N_x V_{ref}}{f_c RC}.$$

Hence

$$\frac{N_x}{N_{max}} = \frac{V_x}{V_{ref}}. \tag{15}$$

Note that the component values R and C and the value of f_c are eliminated by cancellation. Most of the integrated circuits on the market now incorporate features such as autozero and automatic polarity indication.

The dual slope method is particularly valuable in cases where supply frequency interference is a problem, since by arranging for the integrating time (phase I) to include an integral number of supply frequency cycles, the interfering signal is automatically nulled out to zero (figure 12). The conversion time must then be greater than one supply period at the very least. Since the input must be connected for the whole of the integrated period, this method is not suitable for multiplexed inputs except in the case of very slowly varying signals since the scan rate must be very low.

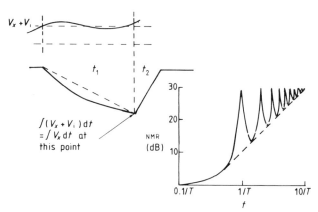

Figure 12. Interference-rejecting property of 'integrating' type of converters.

It is possible to replace the counting logic by a microprocessor. Since decisions are required only at the beginning and end of the phases, the processor is free for other tasks for most of the time.

2.2.2. Frequency analogue methods.

2.2.2.1. Voltage-to-frequency conversion. Instead of using time as an analogue of input voltage, some systems use frequency as an analogue. One advantage of such systems is the ease with which the signal may be transmitted over a twisted pair of wires from a remote location.

The basic principle is again that of charge balance: fixed increments of charge are fed into an integrator at such a frequency that the current due to the unknown voltage is exactly balanced (figure 13). If a clock is used to synchronise the charge balance pulses, the system then becomes a delta–sigma converter (figure 14).

62

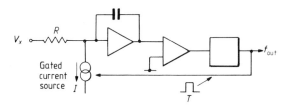

Figure 13. Charge-balance voltage-to-frequency converter. $f_{out}/T = V_x/R$ for charge balance.

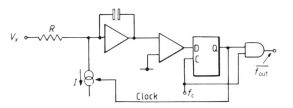

Figure 14. Delta–sigma modulator. $\overline{f_{out}} = f_c V_x/IR$.

The principle is used in some DVM integrated circuits under various names, e.g. 'charge balance', 'quantised feedback'. It should be noted that it is essentially the dual slope method with both phases superimposed, with the discharge time phase II fragmented to fill the whole of phase I (figure 15). It has the same integrating capability for interference reduction.

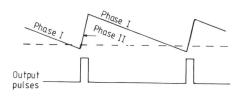

Figure 15. Charge balance viewed as an interleaved dual slope converter.

2.2.2.2. Voltage–frequency/frequency–voltage system. The accuracy of the conversion methods outlined in §2.2.2.1 depends upon the stability and accuracy of the fixed increments of charge which are subtracted from the input current in order to produce a balance. By the addition of a feedback loop the characteristics of the converter may be greatly improved.

Hence, by using a precision frequency-to-voltage feedback element, the forward voltage-to-frequency converter may be given a very much relaxed specification, and a simpler system results.

The feedback circuit uses an analogue switch which connects V_{ref} to the output filter for a fixed time duration T (figure 16). The mean output voltage of the filter is therefore directly proportional to the actuating frequency f_x of the switch.

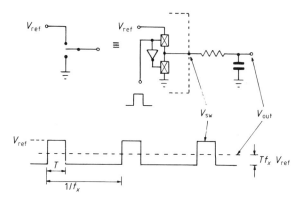

Figure 16. Frequency-to-voltage converter.

The feedback circuit arrangement adjusts the action of the voltage-to-frequency circuit so that $V_{fb} = V_x$; i.e. so that

$$T_0 V_{ref} f_x = V_x. \qquad (16)$$

Thus a measurement of f_x in a gated counter with a gate time T_{ref} will give a digital output

$$N_x = f_x T_{ref} = \frac{V_x}{V_{ref}} \frac{T_{ref}}{T_0}. \qquad (17)$$

By using a common clock to determine both T_{ref} and T_0, their ratio will be made constant, and inaccuracies from these terms can be eliminated, so that $N_x = K(V_x/V_{ref})$, K being a system constant (figure 17). The result is a highly precise system with

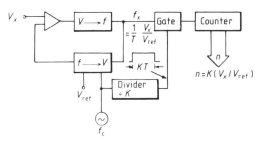

Figure 17. Precision voltage–frequency converter using a frequency-to-voltage feedback element. Time reference $f_c = 1/T$.

very few precision components; a particular feature of this system is the ease with which the conversion circuit may be isolated using pulse transformers.

Note that, since the measurement of frequency is essentially a discrete integration for the period of the gate signal, this method can be given the interference rejection properties of the dual slope converter by choosing the appropriate gate interval.

2.3. Potentiometric methods

2.3.1. Digital-to-analogue converters. The method outlined in § 2.2.2.2 is an example of the use of a precision feedback element to enhance the characteristics of a converter. We will first examine the general design of D–A converters before considering their use in A–D systems.

2.3.1.1. Mark-space converters. In the circuit of figure 16, the mark-space ratio of the switching waveform is converted into a mean voltage in accordance with the expression

$$\overline{V_{\text{out}}} = \frac{T_1}{T_1 + T_2} V_{\text{ref}}. \tag{18}$$

By arranging for $T_1/(T_1 + T_2)$ to be digitally controlled this scheme results in a very simple D–A converter. A typical outline arrangement is shown in figure 18. The counter A runs continuously and produces sequentially all codes from 0 to 2^{M-1}.

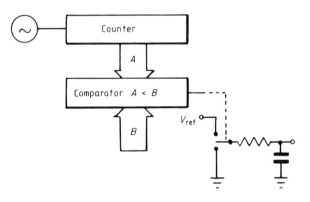

Figure 18. Digital-to-analogue converter using mark-space modulation.

On the 2^Mth count an overflow (carry) occurs and the counter returns to 0; it then continues counting as before. The counter output is a digital version of a ramp waveform, which increases steadily from 0 up to 2^{M-1} and then repeats.

A digital comparator is used to compare the output A with a preset number B. It is easy to see that the digital comparator terminal '$A < B$' is low for all values of A from 0 to $B-1$

inclusive, a total of B clock intervals. Since the cycle time is 2^M intervals, this output has a mean value of $B/2^M$, and by using this to control the output switch, a mean voltage $BV_{\text{ref}}/2^M$ is obtained, $0 \leqslant B \leqslant 2^M - 1$ (figure 19).

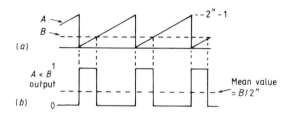

Figure 19. (a), Contents of the counter constitute a digital ramp waveform; (b), output from the comparator.

Smoothing the output requires the filter to have a large time constant (of the order of 2^M times the cycle time), which causes the system to have a long settling time.

The algorithm may be implemented in software, leaving the analogue switches and the filter as the only hardware components required: whether this is a desirable simplification depends on the total use of microprocessor time.

2.3.1.2. Resistor networks. The standard form for the most common type of D–A converter is the $R - 2R$ ladder (figure 20).

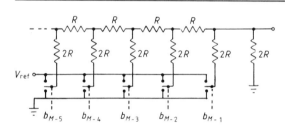

Figure 20. Digital-to-analogue converter using the $R - 2R$ ladder (voltage output).

Here the application of a binary number N_B sets up the appropriate switches, and gives an output from the ladder of

$$V_{\text{out}} = KV_{\text{ref}}(b_{M-1} + b_{M-2}2^{-1} + b_{M-3}2^{-2} \ldots b_0 2^{-M+1}). \tag{19}$$

When b_{M-1} is the MSB (Most Significant Bit), and b_0 the LSB (Least Significant Bit) in an M bit number.

To avoid the need to switch relatively large voltages, the trend of late has been towards inverted ladders, when the ladder is end-fed with V_{ref}, and currents are extracted from the ladder switches by using virtual earth amplifiers (figure 21). In this way the requirements for the voltage handling capability of the switches are relaxed and at the same time more freedom is allowed in the range (and polarity) of V_{ref}. Such converters have the capability of multiplying inputs V_{ref} by a binary number N_B in a general sense, and are termed multiplying DACs.

Recent work has shown that with MOS circuits it is feasible to use the voltage division provided by ratios of capacitor values as a conversion method. Switched capacitor DAC methods are already employed in codecs for telephony, but have as yet made no great impact on the instrumentation field.

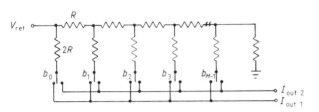

Figure 21. Inverting the ladder network gives a 'multiplying' DAC with current outputs.

2.3.2. A–D converter employing D–A feedback.

2.3.2.1. Digital ramp converters. In this type of converter a counter fed from a clock is connected to the DAC digital input lines. The DAC output therefore steps through all possible values from $0 \times V_{ref}$ to $KN_{B(max)}V_{ref}$ as the counter progresses from a count of 0 to a count of $N_{B(max)}$. A comparator signals the instant when the unknown V_x is attained by the DAC output, and the counter state at that instant is latched and held until another cycle is completed (figure 22).

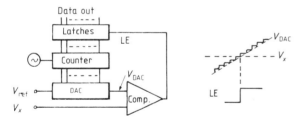

Figure 22. Digital ramp version of the single slope converter.

2.3.2.2. Tracking converters. A modification of this system is to give the counter an up/down capability, and to step the counter on only when there is an error of more than $\pm\frac{1}{2}$ LSB between the value of V_x and the present state of the counter. In this way a tracking ADC results, and, provided the input rate of change is sufficiently low, the counter will always contain the correct representation of V_x (figure 23).

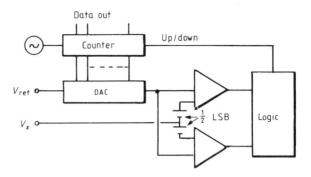

Figure 23. Additional logic and an up/down counter enable the digital ramp converter to track a varying input signal.

2.3.2.3. Successive approximation converter. Counter methods take a long time to attain the initial reading, owing to the need for the counter to count up from zero to the final value in single quantising steps. For high-speed conversion, it is necessary to use some form of successive approximation conversion.

Successive approximation, which is analogous to the 'chemist's balance' method starts with the most significant bit, and proceeds downwards towards the least significant bit, retaining values which were insufficient to cause a balance, and

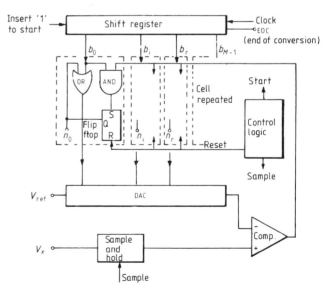

Figure 24. Simplified outline of a successive approximation converter. The logic circuits of a basic cell are usually more subtle in a typical IC.

discarding those which took the sum beyond the balance condition. In the circuit of figure 24 the algorithm would be as shown in figure 25. The advantage of the SA method is the rapidity of conversion. Note that a resolution of 1 part in 2^M is obtained in only M cycles of operation. It is essential, however, for the input voltage to be held to within $\frac{1}{2}$ LSB during the measurement cycle; this normally entails the use of a sample-hold circuit at the input to the converter.

2.4. Multiple comparison methods

In this method, a resistive divider chain produces 2^M different fractions of the reference voltage, where M is the number of bits in the output code. An array of 2^M high-speed voltage comparators is arranged so that each one is fed with its own comparison voltage from the divider chain, the incoming

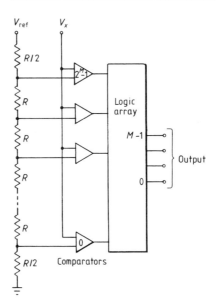

Figure 26. The 'flash' encoder.

analogue signal being connected in common to all the comparators (figure 26). The comparator output signals are encoded by an array of logic elements to give the desired M-bit binary output code. This is a method which was, at one time, fashionable in nucleonics as the 'kicksorter' method. Recent advances in solid state circuit technology have made possible the single chip version, now termed the 'flash encoder'.

The main feature of this approach is the high speed of conversion: a sampling rate of 30 MHz, with 6 or 8 bit output, is claimed by one manufacturer, with an accuracy to within 0.1 dB for frequency components up to 7 MHz.

3. Analogue signal conditioning

For signals which are not directly in voltage form, or are at a level which is incompatible with the dynamic range of the A–D converter, some analogue signal conditioning may be necessary.

In each case the specific form of the input signal and the nature of the source dictate the details of the conditioning circuits, e.g. thermocouples require chopper stabilised amplifiers, pH electrodes require electrometer amplifiers, and so on.

It is usual to incorporate the function of the anti-aliasing prefilter in these circuits.

The fact that the analogue circuits are connected to digital signal processing circuits can often lead to some modifications of the design and can give significant advantages in terms of performance.

For example, an amplifier can have its gain switched in 3 dB steps by signals from a microprocessor (figure 27). In this way, it becomes possible to make an autoranging instrument which

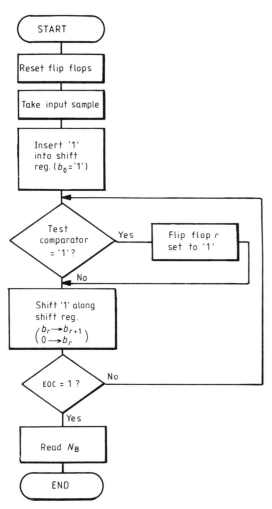

Figure 25. The successive approximation algorithm.

will always arrange its settings so as to make the best use of the dynamic range of the ADC.

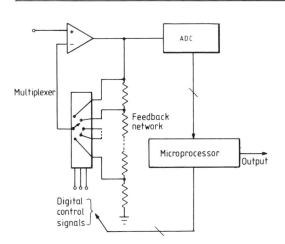

Figure 27. A digitally controlled multiplexer selects the feedback network in 3 dB steps, so that the ADC is always fully exercised by the signal. The output is the ADC signal modified by the gain-setting information contained in the microprocessor.

Consider a 12 bit ADC with a full-scale range of 4.096 volts (1 mV LSB). Although the maximum S/N ratio for sinusoidal signals is 80 dB for a 4V p–p signal, when the input amplitude is 100 mV p–p the S/N ratio is reduced to 48 dB, and is only 28 dB for an input amplitude of 10 mV p–p.

As an alternative we may consider a digitally controlled amplifier with its gain switched in 3 dB steps, followed by an 8 bit ADC. For steady sinusoids, the system could always optimise itself so that the signal to the ADC would never be more than 3 dB below full scale, so that the S/N ratio would always be better than 53 dB. This feature is particularly valuable where it

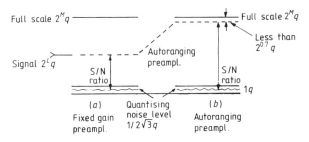

Figure 28. (*a*), Signal-to-noise ratio with a fixed preamplifier; (*b*), improved S/N ratio by autoranging.

is necessary to process small signals superimposed upon a larger signal, if both components are regarded as wanted signals.

The situation is illustrated in figure 28. In (*a*) the *L* bit signal $2^L q$ is compared with the noise level of $q/(2\sqrt{3})$ to give a S/N ratio of $6L + 7.8$ dB, whereas in (*b*) because of the autoranging preamplifier the signal is within 0.5 bit of the full range, and the S/N ratio lies between $6M + 7.8$ dB and $6M + 4.8$ dB.

Conversely, let us consider the case where the signal consists of a large-amplitude unwanted component (such as interfering hum) superimposed upon a lower-level wanted signal. We decide to remove the unwanted signal by digital filtering after the ADC process. The new situation is illustrated in figure 29.

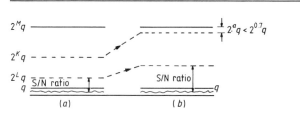

Figure 29. A large interfering signal prevents the full benefits of autoranging from being obtained.

The signal levels for large and small components are now the same for both (*a*) and (*b*), but in (*b*) autoranging has taken place so that the large signal is within 0.5 bit of full scale. Here the wanted signal component is K–L bits below M–a where a is less than 0.5 bit; hence the S/N ratio lies between $6(M + L - K) + 4.8$ and $6(M + L - K) + 7.8$ dB. This compares with $6L + 7.8$ dB in (*a*). Thus the advantage of the autoranging circuit has been reduced to the extent of $6(M - K)$ dB because of the signal $2^K q$.

It is apparent that the full advantage of the autoranging circuit can be realised only if unwanted signals are removed *prior* to conversion; that is, the analogue signal conditioning must include analogue filtering to remove the unwanted signals.

These considerations apply not only to autoranging circuits but also to cases where fixed-gain amplifiers are employed, and it is necessary to avoid the expense of using high-resolution converters, where a lower resolution would be acceptable if the unwanted signals were not present.

A typical case is the measurement of a low-level pilot tone superimposed on a communication channel. The analogue prefiltering in this case must be capable of extracting the pilot tone and suppressing all other signals before analogue–digital conversion.

4. Processing of signals in the digital domain

When analogue signals have been translated into the digital domain, the result may be regarded as time series of data.

Analogue operations on the analogue data may be simulated by discrete operations on the time series data, using the technique of numerical analysis, often in simplified form.

For example, the integration operation

$$I = \int_{T_1}^{T_2} x(t)\, dt \qquad (20)$$

may be approximated by the summation operation

$$I \simeq T \sum_{n_1}^{n_2} x(nT) \qquad (21)$$

where T is the interval between samples, and $n_1 T = T_1$, and $n_2 T = T_2$. Similarly, the differentiation operation

$$x'(x) = \frac{dx(t)}{dt} \qquad (22)$$

can sometimes be approximated by

$$x'(nt) \simeq [x((n+1)T) - x(nT)]/T, \qquad (23)$$

although here quantising noise often invalidates the method.

4.1. Mathematical descriptions
It is necessary to have the appropriate mathematical techniques to describe the digital domain version of the signal.

In the analogue domain Fourier and Laplace transforms are used to relate the frequency and time characteristics of a signal. Thus

$$X_a(s) = \int_0^\infty x_a(t) \exp(-st)\, dt \qquad \text{(Laplace)} \qquad (24a)$$

$$X_a(j\omega) = \int_{-\infty}^\infty x_a(t) \exp(-j\omega t)\, dt \qquad \text{(Fourier)} \qquad (24b)$$

4.1.1. Discrete Fourier transform. We may regard $\tilde{x}(t)$, the sampled version of $x_a(t)$, as the product of $x_a(t)$ with a function $d(t)$ representing a train of unit impulses (figure 30)

$$d(t) = \sum_{n=-\infty}^\infty \delta(t - nT). \qquad (25)$$

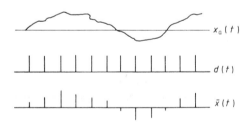

Figure 30. Impulse train $d(t)$ multiplied by $x_a(t)$ to produce the sampled version $\tilde{x}(t)$.

Thus

$$\tilde{x}(t) = \sum_{m=-\infty}^\infty x_a(t)\delta(t - nT) \qquad (26a)$$

$$= \sum_{n=-\infty}^\infty x_a(nT)\delta(t - nT), \quad \text{(since the product exists} \qquad (26b)$$
$$\text{only at the instants}$$
$$t = nT).$$

The Fourier transform of $\tilde{x}(t)$ is

$$\bar{X}(j\omega) = \int_{-\infty}^\infty \tilde{x}(t) \exp(-j\omega t)\, dt \qquad (27)$$

$$= \sum_{n=-\infty}^\infty x(nT) \int_{-\infty}^\infty \delta(t - nT) \exp(-j\omega t)\, dt \qquad (28)$$

$$= \sum_n x(nT) \exp(-j\omega nT). \qquad (29)$$

Apart from a scale factor T, this could have been obtained from (24b) by applying the discrete approximation of (21).

By convoluting the transforms of $d(t)$ and $X_a(t)$, we can arrive at $\bar{X}(j\omega)$ by another route.

$$D(j\omega) = \int_{-\infty}^\infty d(t) \exp(-j\omega t)\, dt. \qquad (30a)$$

By expanding $d(t)$ as a Fourier *series*, and then performing the Fourier transform, the result is

$$D(j\omega) = \frac{2\pi}{T} \sum_{n=-\infty}^\infty \delta\left(\omega - \frac{2\pi n}{T}\right). \qquad (30b)$$

Thus

$$\bar{X}(j\omega) = \frac{1}{2\pi} \int_{-\infty}^\infty X_a(j\xi)D(j(\omega - \xi))\, d\xi \qquad (31a)$$

$$= \frac{1}{T} \sum_{-\infty}^\infty X_a\left(j\left(\omega - \frac{2\pi n}{T}\right)\right). \qquad (31b)$$

This shows that the Fourier transform of a sampled signal is the summation of shifted versions of the original Fourier transform (figure 31(a)). It is now clear why prefiltering is required; if the original spectrum $X_a(j\omega)$ was too wide the shifted versions would overlap (figure 31(b)). $\bar{X}(j\omega)$ is known as the Discrete Fourier Transform of $\tilde{x}(nT)$.

In practice a finite number of samples is available, and the DFT is then written

$$\bar{X}(j\omega) = \sum_{n=0}^{N-1} x(nT) \exp(-j\omega nT). \qquad (32)$$

Because the number of samples is limited, it is no longer possible to compute $\bar{X}(j\omega)$ for all values of ω; only N discrete frequency points can be calculated from N data inputs, at values $\omega = 2\pi m/NT$, $m = 0 \ldots N-1$.

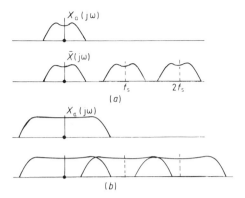

Figure 31. (a), Sampling reproduces $X_a(j\omega)$ at multiples of f_s; (b), if f_s is too low the sampled spectra overlap.

The DFT then reduces to

$$\bar{X}(m) = \sum_{n=0}^{N-1} x(n) \exp(-j2\pi mn/N). \qquad (33)$$

The Inverse DFT is given by

$$\bar{x}(n) = (1/N) \sum_{n=0}^{N-1} \bar{X}(m) \exp(j2\pi mn/N) \qquad (34)$$

4.1.2. z-transform. The Laplace transform can also be computed for the sampled waveform. Starting from

$$\bar{x}(t) = \sum_{n=0}^{\infty} x_a(nT)\delta(t-nT) \qquad (26b)$$

we use the Laplace delay operator to obtain the transform

$$\mathscr{L}\delta(t-nT) = \exp(-snT). \qquad (35)$$

Hence

$$\bar{X}(s) = \sum_{n=0}^{\infty} x(nT) \exp(-snT) \qquad (36)$$

represents the Laplace transform of the sampled waveform.

By making the substitution

$$\exp(-sT) = z^{-1} \text{ (unit delay)} \qquad (37)$$

this may be rewritten

$$\tilde{X}(z) = \sum_{n=0}^{\infty} x(n)z^{-n}. \qquad (38)$$

This is termed the z-transform of $\bar{x}(t)$. It is easily shown that $\bar{X}(j\omega)$ and $\tilde{X}(z)$ are equivalent, where s is constrained to be on the frequency axis, i.e. $s = j\omega$.

4.2. Digital filter descriptions

These relationships between sampled time domain functions and their frequency domain representations provides the basis for the design of digital filters. Thus a network with an impulse response $h(n)$ has a DFT $\bar{H}(j\omega)$ and a z-transform $\tilde{H}(z)$.

As in the case of continuous signals, frequency filtering may be expressed as the product of the Fourier transform and the network response. Thus, using the DFT, a network $\bar{H}(j\omega)$ with an input $\bar{X}(j\omega)$ has an output

$$\bar{Y}(j\omega) = \bar{H}(j\omega) \cdot \bar{X}(j\omega); \qquad (39)$$

similar considerations apply to the z-transform.

Using discrete convolution, the inverse z-transform of $\tilde{X}(z)$ operated upon by $\tilde{H}(z)$ is

$$\bar{y}(n) = \sum_{k=0}^{N-1} h(k)x(n-k). \qquad (40)$$

In terms of discrete mathematics, this represents the summation of the product of terms $h(k)$ with delayed versions of samples of x taken at times previous to nT. This operation could be represented as a series of delay elements operating upon $\bar{x}$, with multiplications by factors h followed by a summation (figure 32).

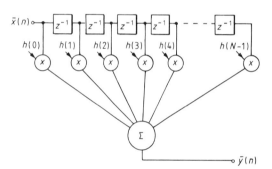

Figure 32. Basic nonrecursive structure to implement the algorithm

$$y(n) = \sum_{K=0}^{N-1} x(n-K)h(K)$$

Such a 'structure' may be regarded not only as a software algorithm for performing the computation described by the z-transform, but also as an outline circuit diagram for an instrument which performs the same function.

4.3. Recursive and nonrecursive structures

The 'structure' which performs the basic z-transform (or DFT) is termed a nonrecursive filter (there is no feedback).

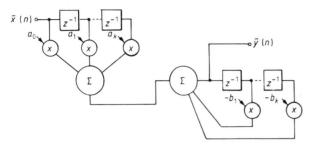

Figure 33. A recursive structure added to a nonrecursive structure. The two adders may be combined as one unit.

A simple extension to the algorithm involves operating on the output of the filter and feeding the result back upon itself (figure 33). The expression is then

$$y(n) = \sum a(k)x(n-k) - \sum b(k)y(n-k). \quad (41)$$

This results in a response

$$\tilde{H}(z) = \tilde{A}(z)/\tilde{B}(z) \quad (42)$$

or

$$\bar{H}(j\omega) = \bar{A}(j\omega)/\bar{B}(j\omega). \quad (43)$$

Such a structure is termed recursive. It is obvious that a recursive structure may be made up using two nonrecursive structures.

In addition to performing filtering, the same general structure may be used to carry out other functions: for example the 'structure' of figure (34) shows a cross-correlator.

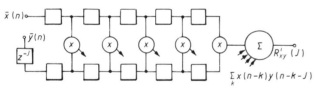

Figure 34. A cross-correlator structure which compares $\tilde{x}(n)$ with $\bar{y}(n-j)$.

It is not intended here to discuss digital filter design, as this is covered very fully in the literature. The examples given below, however, are included to illustrate some of the features of the various forms of filter structure.

4.3.1. Nonrecursive structure
A 'moving average' filter, also known as 'smoothing by fives' is often used for smoothing data. It has the algorithm

$$y(n) = \tfrac{1}{5}[x(n-2) + x(n-1) + x(n) + x(n+1) + x(n+2)]$$
$$= \tfrac{1}{5}[x(n)z^{-2} + x(n)z^{-1} + x(n) + x(n)z + x(n)z^2]$$
$$= \tfrac{1}{5}x(n)(z^{-2} + z^{-1} + z^0 + z + z^2). \quad (44)$$

Thus

$$\tilde{H}(z) = \tfrac{1}{5} \sum_{r=-2}^{2} z^r \quad (45)$$

substituting $\exp(j\omega T)$ for z we obtain

$$\bar{H}(j\omega) = \tfrac{1}{5} \sum_{r=-2}^{2} \exp(rj\omega T) \quad (46a)$$
$$= \tfrac{2}{5} \cos 2\omega T + \tfrac{2}{5} \cos \omega T + \tfrac{1}{5}. \quad (46b)$$

This structure is shown in figure 35 and its response is plotted in figure (36(a)). It can be seen that this has a low-pass characteristic, but has a significant response at half the sampling frequency.

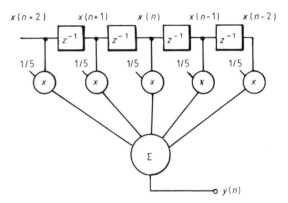

Figure 35. The 'smoothing by 5's' structure for data smoothing. In a real-time circuit $x(n+2)$ is not available since it represents advance information, and the input will be $x(n)$, the output then being $y(n-2)$.

If the outer weights are halved, and the gain adjusted accordingly

$$\tilde{H}(Z) = \tfrac{1}{4}(\tfrac{1}{2}z^{-2} + z^{-1} + z^0 + 1z^1 + \tfrac{1}{2}z^2) \quad (47a)$$

and

$$\bar{H}(j\omega) = \tfrac{1}{4}(\cos 2\omega T + 2 \cos \omega T + 1). \quad (47b)$$

This response is shown in figure (36(b)). The improvement in the response by modifying the values of the outer coefficients is termed 'windowing', of which the above is a simple example.

It should be noted that the quantising noise on $\tilde{x}(n)$ tends to be averaged by the summations, and $\bar{y}(n)$ will have a better S/N ratio than $\tilde{x}(n)$.

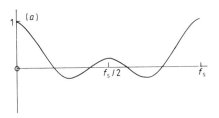

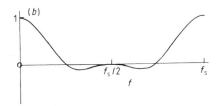

Figure 36. (a), Frequency response of the 'smoothing by 5's' structure; (b), reduction of ripple at $f_s/2$ by adjusting the outer coefficients: a simple illustration of 'windowing'.

4.3.2. Recursive structure. A simple algorithm is given by

$$y(n) = x(n) + \tfrac{1}{2}y(n-1) \qquad \text{(figure 37(a)).} \qquad (48)$$

This gives

$$\tilde{H}(z) = 1/(1 - \tfrac{1}{2}z^{-1}). \qquad (49)$$

From this we obtain

$$\bar{H}(\mathrm{j}\omega) = (1 - \tfrac{1}{2}\exp(-\mathrm{j}\omega T))^{-1} \qquad (50)$$

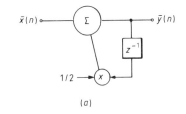

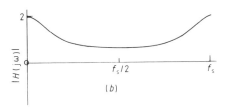

Figure 37. First-order recursive structure with a low-pass response.

which gives the modulus

$$|\bar{H}(\mathrm{j}\omega)| = (1\tfrac{1}{4} - \cos \omega T)^{-1/2}. \qquad (51)$$

This is plotted in figure 37. Note that a single delay element and coefficient have given a much sharper low-pass response than would be available with an equivalent nonrecursive structure. A simple modification to the algorithm would be to add a nonrecursive term (figure 38(a)).

$$y(n) = x(n) + x(n-1) + \tfrac{1}{2}y(n-1). \qquad (52)$$

This then gives

$$\tilde{H}(z) = \frac{1 + z^{-1}}{1 - \tfrac{1}{2}z^{-1}}, \qquad (53)$$

resulting in the frequency response

$$|\bar{H}(\mathrm{j}\omega)| = |\cos \omega T/2|/\sqrt{1\tfrac{1}{4} - \cos \omega T}. \qquad (54)$$

This is shown in figure 38(b). The result is now a rather sharper low-pass filter with a null in the response at half the sampling frequency.

The form of response $\tilde{H}(z)$ may be investigated by plotting the poles and zeros of $H(s)$ on the s plane.

This technique is fully covered in the literature; here it is sufficient to note that the vertical axis of the s plane transforms to a unit circle on the z plane, with the s plane left-hand half being enclosed within the circle (figure 39). It follows that z plane poles outside this circle give an unstable system.

In cases where z plane poles are required to be only just within the circle in order to give high Q resonances, it may be difficult to define them adequately because of the inadequate resolution of the digital representation of the coefficients.

The problem does not arise with nonrecursive structures

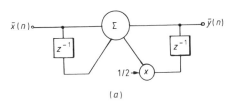

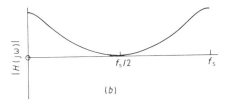

Figure 38. Addition of first-order nonrecursive stage to eliminate response at $f_s/2$.

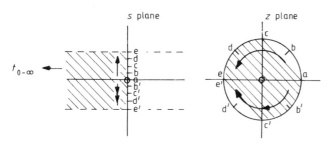

Figure 39. Mapping of s plane onto the z plane. The shaded strip to $-\infty$ maps into the unit circle, and the corresponding strip to $+\infty$ occupies the rest of the z plane. Points e and e' represent the half sampling frequency $\pm f_s/2$; further parallel strips of the s plane also map into the *same* regions of the z plane. (See for example the products of sampling f_x in figure 4.)

because these are all zero networks which are not so sensitive to the exact positioning of any particular singularity.

4.4. Choice of structure

The choice of structure depends upon many factors. Nonrecursive structures are unconditionally stable, and can be designed for any given frequency response using the DFT. However, to obtain a good approximation to the design specification requires many stages and may demand very high-speed operations, as will be shown in the next section. They are best suited to designs where modest slopes are called for in the frequency response.

Recursive structures need fewer operations of delay, multiplication and addition, and are thus potentially much faster in operation. They can give very sharp resonant peaks because of the poles generated by the recursion. However, they can be unstable due to round-off errors, and their design must be carried out with extreme care.

5. Hardware design

"... digital hardware design, like most engineering, is based on the intuitive attitude of the designers to a large number of ill-defined tradeoffs". This view was expressed by Rabiner and Gold in their book, '*Theory and Application of Digital Signal Processing*'.

The nonrecursive filter structure of figure 32 will be considered as an example: extensions to the more complex structures are relatively straightforward. The 'structure' can be viewed as a software algorithm or as an outline for hardware implementation.

5.1. Multiple operations

Consideration of the software case immediately introduces the concept of multiplexed operation since it is obvious that a computer can only add two numbers at a time, and also that only one multiplication algorithm can be run at one time. Thus the N multiplication operations and subsequent N-input addition must be performed in series. If the system is to operate in 'real' time, then the whole sequence of operations must be completed within the period of the sampled data. The multiplexing concept

Table 2. States during one complete time period. For an N-stage structure, the basic time period T must be subdivided into N phases $\varphi_1 - \varphi_N$ and the clock period is T/N.

Time	Input switch	Input	x_0	x_1	x_2	H_0	H_1	H_2	H_3	Product	Sum
$t = nT$											
φ_1	Recirc.	$x(n)$	$x(n-1)$	$x(n-2)$	$x(n-3)$	a_0	a_1	a_2	a_3	$a_3 x(n-3)$	$a_3 x(n-3)$
φ_2	Load		$x(n)$	$x(n-1)$	$x(n-2)$	a_3	a_0	a_1	a_2	$a_2 x(n-2)$	$\left\{ \begin{array}{l} a_3 x(n-3) \\ +a_2 x(n-2) \end{array} \right.$
φ_3	Recirc.		$x(n-2)$	$x(n)$	$x(n-1)$	a_2	a_3	a_0	a_1	$a_1 x(n-1)$	$\left\{ \begin{array}{l} a_3 x(n-3) \\ +a_2 x(n-2) \\ +a_1 x(n-1) \end{array} \right.$
φ_4	Recirc.		$x(n-1)$	$x(n-2)$	$x(n)$	a_1	a_2	a_3	a_0	$a_0 x(n)$	$\left\{ \begin{array}{l} a_3 x(n-3) \\ +a_2 x(n-2) \\ +a_1 x(n-1) \\ +a_0 x(n) \end{array} \right\} = y(n)$
$t = (n+1)T$											
φ_1	Recirc.	$x(n+1)$	$x(n)$	$x(n-1)$	$x(n-2)$	a_0	a_1	a_2	a_3	$a_3 x(n-2)$	$a_3 x(n-2)$
φ_2	Load										

is also valuable when hardware designs are being considered, since this gives tremendous savings in hardware provided the circuits employed can operate at a high enough rate. When multiplexing techniques are applied, there is also a requirement for the storage and retrieval of N successive samples of $x(n)$ and $h(n)$ in addition to the multiplication and summation functions.

5.2. 'Think Analogue, Build Digital'
It is often easier to envisage the operation of a system by considering an equivalent hardware realisation, even to the extent of giving the latter an 'analogue' personality. Figure 40(a) and (b) show conceptual equivalents. It is assumed that the filter coefficients have somehow been introduced into the coefficient delay line, where they circulate indefinitely. (The circulation time in the analogue case defines the sampling period.)

For $N=4$ in figure 32 the states in the digital equivalent (figure 40(b)) are shown in table 2. Note that the coefficient shift register requires one more stage than the X register, and also that during the load cycle a new sample of $x(n)$ is introduced into the X register whilst the oldest sample $x(n-3)$ is lost.

Such a system is directly realisable in hardware form, using an array of M shift registers in parallel to deal with M-bit words (figure 41), although such a system is rather cumbersome.

5.3. Serial arithmetic
Many workers have chosen to simplify the problem by using a serial representation; a single shift register of appropriate length is then required. The multiplier can now be a shift-and-add type, with the x data being entered serially, but the coefficient bits must still be in parallel form (figure 42). Normally the data will be in 2's complement form, LSB first[†]. As an alternative, fast 'bit-slice' chips such as the AM2903, could be microprogrammed as a shift-and-add multiplier.

An example of a basic filter section is shown in figure 43. Here four multipliers have been used to realise a second-order recursive filter.

A cascade of filter sections can be realised with very little hardware penalty by multiplexing the filter coefficients and adding an extra shift register to store the partial result (figure 44). On the first iteration, coefficients $h_1(r)$ are read out from

memory to implement § 1, followed by the § 2 coefficients $h_2(r)$ for the second iteration, and so on. In this way the same basic hardware for a single section is used over and over to form a high-order filter.

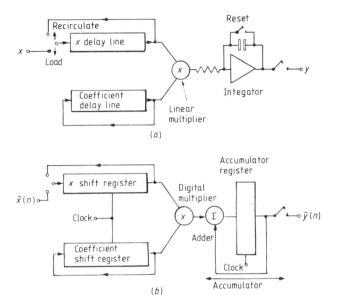

Figure 40. (a), Analogue concept of a nonrecursive structure; (b), digital shift register equivalent.

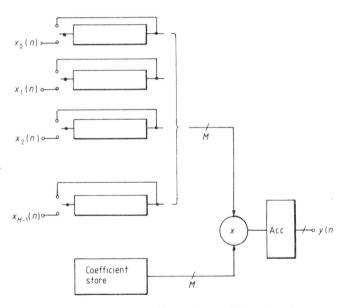

Figure 41. An M-bit input $x(n)$ requires M shift registers in parallel.

[†] Numbers expressed in natural binary form are always positive. In two's complement notation the most significant bit of the number is treated as a sign bit, (0 for positive, 1 for negative); for a positive number the remaining bits represent the number directly, and for a negative number less significant bits can be formed by subtracting its modulus from 2^M. For example 0111 represents $+7$, 0001 is 1 (both positive), 1111 represents -1, and 1000 is -8. In general the range of an M bit number is from $2^{M-1}-1$ to $-(2^{M-1})$. By using a serial representation, commencing with the least significant bit, it becomes easy to double a number by delaying it.

5.4. The Peled–Liu technique

Having simplified the storage problem by serial representation, the multiplication problem remains, and the shift-and-add technique still requires a fair amount of hardware.

Peled and Liu have shown that the shift-and-add multiplier may be replaced by a Read Only Memory (ROM) and an accumulator, as shown in figure 45. The ROM contains all possible partial products of bits from x and y multiplied by bits from coefficients a and b.

This design is based upon the fact that the ROM has only 5 inputs, so that for a given filter characteristic there are only $2^5 = 32$ possible outputs. Data exist serially in SR1 and SR2, SR3 and SR4, and in parallel in PR5 and PR6.

As serial data is clocked through the registers a result is accumulated in PR6; at the end of the operation the contents of PR6 are loaded in parallel into SR3 in readiness for the next sample period.

This basic section can also be cascaded upon itself by multiplexing (figure 46). Since the ROM contents depend upon

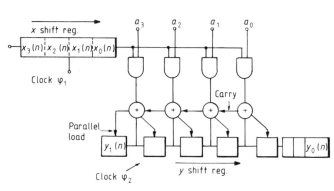

Figure 42. The shift-and-add method of multiplication. The multiplier a is supplied in parallel, and the multiplicand arrives in serial form. The partial products are added in parallel into the product shift register; the final output may be taken in parallel or serial form as required. The X states are shown at the start of the operation, the Y states at the end.

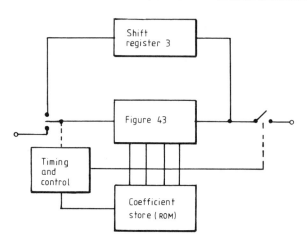

Figure 44. A recirculating scheme enabling the same circuit to perform cascaded filtering by changing the filter coefficient at each circulation.

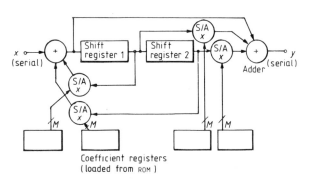

Figure 43. Use of shift-and-add multipliers to construct a digital filter where the input and output are in serial form; the filter coefficients are in parallel form.

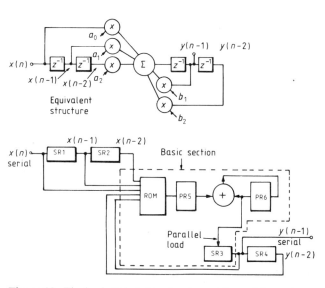

Figure 45. The basic Peled–Liu structure. All possible products have been pre-computed and are held in the ROM, thus eliminating the need for multipliers. SR, serial shift register; PR, parallel register.

the coefficients for each section, a different segment of ROM is called up as each section is simulated. Note that a set of output shift registers is required for each 'section' to form input shift registers for the next 'section'.

The main disadvantage of the Peled and Liu approach is the need to reprogram the ROM if new filter coefficients are required. With the 'multiplier' approach, previously described, the coefficients may be changed at any time.

5.5. *Parallel arithmetic*

Whilst the serial methods described above have given us high-speed operation with reduced hardware, the advances in silicon technology which have made the microprocessor possible have also made parallel operation much more feasible.

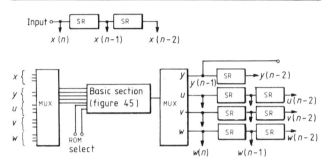

Figure 46. Recirculating the Peled–Liu structure to obtain a multi-section filter. A separate area of ROM is used at each pass.

Filter section	ROM select	Input select	Output feed
1	00	$x, w(n), w(n-1)$	w
2	01	$w, v(n), v(n-1)$	v
3	10	$v, u(n), u(n-1)$	u
4	11	$u, y(n-1), y(n-2)$	y

With Random Access Memory (RAM) and ROM readily available, providing direct storage of M-bit wide words, it has become more convenient to simulate shift register operation by address manipulation. Thus the coefficient shift register in figure 40(b) is replaced by ROM with sequential addressing, i.e. instead of moving data along from one shift register stage to the next, the data remain stationary and are accessed by a moving pointer (figure 47(a)). The design of the x register pointer, which has to simulate the shifting-out of ancient data and the insertion of new data in the correct sequence, is rather more complicated. One solution is to derive the address pointers from two counters, modulo N and modulo $N-1$. The modulo N counter drives the coefficient ROM and provides the system timing; the modulo $N-1$ counter drives the X store which is a RAM. The new data $x(n)$ are

written into the X RAM under the command of the modulo N counter into whatever address is supplied by the modulo $N-1$ counter at that time.

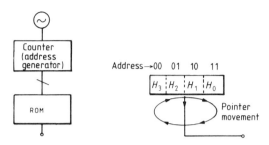

Figure 47. Simulating the coefficient shift register by using a ROM and a moving address pointer.

The outputs of the ROM and RAM are multiplied and the products are accumulated under the control of the modulo N counter. To complement the relative simplicity of the RAM/ROM/moving-pointer approach the multiplication and accumulation operation can now be performed at high speed using a family of single-chip 'MAC' circuits from TRW. A 'MAC' can be used as the kernel of an Arithmetic Logic Unit (ALU) to form a medium-speed signal-processing system (figure 48) which simulates the basic structure of figure 32. By arranging for the y output to be fed back into the RAM, the structure of figure 33 can also be simulated.

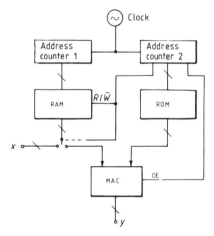

Figure 48. A nonrecursive filter using moving pointer techniques to simulate shift registers, and a multiplier-and-accumulator chip, in a direct emulation of figure 40(b).

Note that this structure is now very close to that of a general-purpose microcomputer. Another alternative is to use fast bit-slice chips to construct a microprogrammed MAC. The structure then is even more closely akin to a microcomputer.

5.6. Signal processors in software

It is now appropriate to examine the possibility of using a microcomputer as a signal processor. For virtually any microprocessor it is found that the generation of 'shift register' pointer addresses is a trivial programming problem; the main difficulty lies with the multiply function. The designer has then to choose whether to use a software 'multiply' routine, a ROM look-up table, or a high-speed multiplier chip. This is where the economics of the day rules the final selection.

A flow chart for a hypothetical microprocessor nonrecursive filter program is shown in figure 49. It could equally well be a program flowchart for data smoothing on a mainframe computer, without, of course, the 'real time' feature of the microprocessor application.

5.7. Operating speeds

Having reviewed the main realisation techniques, we are now in a position to examine the speed possibilities of each approach.

For the shift-and-add multiplier, an $M \times M$ bit multiplication takes $t_m = M(M-1)t_c + Mt_s + Mt_r$, where t_c is the carry propagation time through a 1-bit adder, t_s is the sum propagation time, and t_r is the x register settling time. The configuration of figure 32, with N stages of shift register, will therefore take Nt_m to process each sample, and the maximum sampling rate is $1/Nt_m$. If the filter section is cascade-multiplexed to form K sections, the maximum sampling rate reduces to $1/KNt_m$. Assuming the use of a high-speed logic family with $t_c = t_s = t_r = 2$ ns, then for an 8-bit word length $t_m = 144$ ns. Using the configuration of figure 43, the forward and feedback paths are computed at the same time and do not hinder each other. For a second-order section $N = 2$. If we assume that an eighth-order filter is being synthesised then $K = 4$. The maximum sampling rate is therefore 0.87 MHz. For the Peled–Liu realisation the maximum rate is set either by the access time of the ROM, t_A or the addition time for two m-bit numbers, t_d, since both operations occur concurrently. Typically $t_A = 50$ ns for a bipolar memory and, using the same logic as before, $t_d = 18$ ns. The maximum bit rate is thus 20 MHz. If 8-bit words are assumed, then the word rate (i.e. the sample rate)

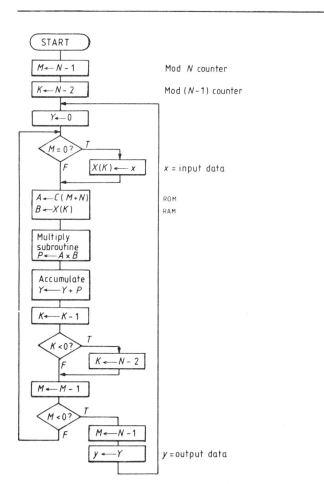

Figure 49. Algorithm for a software nonrecursive filter.

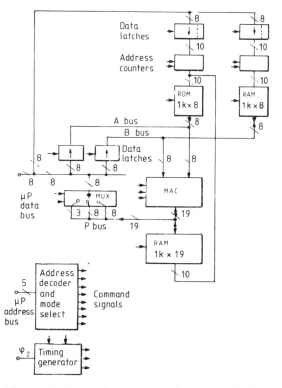

Figure 50. A general-purpose signal processor designed as a microprocessor peripheral.

is thus limited to 2.5 MHz, and if four cascaded sections are simulated the maximum sampling rate reduces to 0.63 MHz. Note however, that since the memory access time is the limiting factor, another logic family with slower speed and lower dissipation could be selected. Peled and Liu show that their basic filter structures may be combined in parallel to give higher sampling rate: the basic second-order section could then operate at 20 MHz *word* rate by parallelling eight circuits; the four-section cascade-multiplexed filter could then operate at 5 MHz.

Turning to the parallel realisation using the pointer 'shift register' and a 'MAC' chip, we find that memory access time t_A is 50 ns as before, and that the multiply-accumulate time t_{MAC} is 100 ns (max) for an 8×8 bit chip. The operating rate is now limited by t_{MAC}. For an N stage filter the sampling rate is thus $1/Nt_{MAC}$. By appropriate programming of the control signals a second-order recursive section ($N = 2$) will thus have a sampling rate of 5 MHz, which will of course be reduced proportionately if cascade-multiplex operation is carried out. In comparison, the operating speed of the software system will be very low indeed, by several orders of magnitude, and such a system is suitable only for very low-bandwidth signals.

5.8. A practical example
Combinations of microprocessors with special-purpose digital hardware provide a sensible halfway house for many applications.

Figure 50 shows how a signal-processing board using a MAC may be incorporated in a microcomputer system. Here the microcomputer provides command signals which define the function of the signal processor, and also feeds in data (the raw data rate being within the capability of the micro).

Whilst the signals are being processed, the microcomputer is free to undertake other tasks. Finally, processed data may be extracted by the micro to be used in other parts of the system, formatted for display, etc.

For simplicity the MAC is run at the microcomputer system clock rate (1 MHz). It is capable, for example, of performing real-time correlation (256 points) between three channels of data sampled at 1 kHz rate, and a fourth (reference) channel.

The board can be set up to perform the following signal-processing functions:

Cross correlation
Autocorrelation
Nonrecursive filtering
Recursive filtering
Signal averaging
Probability density.

In addition the multiplier is accessible to the microcomputer for use in fast Fourier transformation.

6. Conclusions
This article has attempted to pick out some of the major features in a fast-developing topic which has become much easier to implement with recent advances in digital integrated circuits in conjunction with microprocessors.

No attempt has been made to discuss the details of recursive filter design using various transformations, windowing techniques in nonrecursive filters, or the effects of round-off and truncation both on coefficients and on data. The topic of digital waveform generation has also been omitted. All these aspects are well covered in the literature, of which a representative list will be found in the references below.

Having designed his algorithmic structure, the designer's task is then to find an optimum means of performing that algorithm. The optimum solution changes as the state of the art progresses. There are some very subtle tricks which can be applied when the ultimate in performance is being aimed at, and this area of design is one of the most challenging in the whole field of modern electronics.

Further reading
Data conversion
Subskind A K (ed.) 1967 *Notes on Analog–Digital Conversion Techniques* (Boston: MIT Press)

Schmid H 1970 *Electronic Analog/Digital Conversions* (New York: Van Nostrand Reinhold)

Dooley D J (ed.) *Data Conversion Integrated Circuits* (London: IEEE)

Noise analysis: signal analysis
Bendat J S and Piersol A G *Random Data: Analysis and Measurement Procedures* (London: Wiley)

Oliver B M and Cage J M *Electronic Measurements and Instrumentation* (New York: McGraw Hill)

Papoulis A 1977 *Signal Analysis* (New York: McGraw Hill)

Kuo B C 1962 *Automatic Control Systems* (London: Prentice Hall). (A good account of *z* transform theory).

Digital signal processing
Rabiner L R and Gold B 1975 *Theory and Application of Digital Signal Processing* (Englewood Cliffs, New Jersey: Prentice Hall)

Hamming R W 1977 *Digital Filters* (Englewood Cliffs, New Jersey: Prentice Hall)

Steorns S D 1975 *Digital Signal Analysis* (Rochelle Park, New Jersey: Hayden)

Gold B and Rader M 1969 *Digital Processing of Signals* (New York: McGraw Hill)

Peled A and Liu B 1976 *Digital Signal Processing* (New York: Wiley)

Oppenheim A V 1978 *Applications of Digital Signal Processing* (Englewood Cliffs, New Jersey: Prentice Hall)

Hardware
TRW LSI Products, PO Box 2472, La Jolla, California, USA

AMI Microsystems Ltd, Princes House, Princes Street, Swindon, UK

Advanced Micro Devices, 901 Thompson Place, Sunnyvale, California, USA

Chapter 7

Advances in lock-in amplifiers

M L Meade

Faculty of Technology, The Open University, Walton Hall,
Milton Keynes MK7 6AA, UK

Abstract. The key specification points of lock-in amplifier systems for signal recovery and signal characterisation are reviewed and it is shown how these can be improved and modified by more advanced system design. The configurations of several commercially available systems are described and the facilities available in computer-controlled lock-in systems are discussed briefly, together with some new application areas.

1. Introduction

A major step in the design of any experiment is to arrange for a distinctive time variation which disassociates the output signals from random disturbances. This can be achieved by imposing some structure on the excitation source and physical limitations in the design of the experiment often mean that *periodic* excitations must be used. This explains the abundance of mechanical and electromechanical 'choppers' deployed in a wide range of applications to interrupt normally 'static' excitations.

In many experiments the output signal appears at the fundamental excitation frequency or bears a harmonic relationship to it. In either case the relative phase of the signal or its amplitude might be of interest and measurements will be impeded by the presence of noise and interference.

A signal recovery 'problem' is acknowledged to exist when adequate separation of signal and noise cannot be achieved by the use of linear filters at the output of the experiment. In this event a true signal recovery system must be used to monitor the variations in amplitude and phase of the signal. By this we mean a system which takes specific account of the 'character' of the signal imposed by the choice of excitation source so as to identify it against a noise background.

It is this aspect of signal recovery which involves the use of lock-in amplifiers as high-performance demodulators capable of operating under conditions of adverse signal-to-noise ratio.

To many research workers the terms 'signal recovery' and 'lock-in recovery' are virtually interchangeable, a fact which reflects the extreme importance of lock-in systems in all fields of scientific investigation. To concentrate solely on signal recovery applications, however, would deny the importance of lock-in systems in the *precision* measurement and characterisation of relatively 'clean' signals. This is a relatively new application area made possible by advances in system and circuit design during the last few years.

In order to trace some of these developments it must be assumed that the reader is familiar with at least the basic principles of lock-in detection as described in the literature (Hieftje 1972a, b, Blair and Sydenham 1975, Fisher 1977). These will be reviewed briefly but the first main objective will be to clarify some of the key specifications used by leading lock-in amplifier manufacturers. This will provide a basis for reviewing the various methods which have been adopted in commercial systems to improve and modify the performance of 'conventional' systems for use in more demanding applications.

The standard literature is found to be very disappointing in this respect. Although lock-in or synchronous detection techniques are well-known to authors in telecommunications (for example Taub and Schilling 1971) critical aspects of performance such as the provision of adequate dynamic range are rarely examined at all. Occasionally, papers appear on state-of-the-art developments in circuit design (Carter and Faulkner 1977 a, b) but for the most part the information here is taken from manufacturers' literature, notably from Brookdeal Electronics, Princeton Applied Research Corporation and Ithaco (Munroe 1973).

Blair and Sydenham (1975) have given an indication of the range of application of lock-in techniques. In this review attention will be confined to areas which have recently benefited from the availability of computer-controlled equipment.

2. Essential principles of lock-in systems

Experiments exploiting the benefits of lock-in detection almost invariably use a *reference* signal derived from the excitation source as shown in figure 1.

Lock-in systems are sometimes described as special purpose *correlators*. The final response depends on the high degree of correlation which is known to exist between the signal of interest and the reference. The presence of correlation is tested in the usual way by multiplying together these two inputs to form the product:

$$v_p(t) = r(t)(s(t) + n(t)).$$

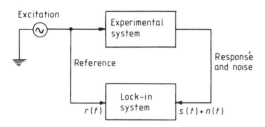

Figure 1. Basic arrangement of an experimental system with lock-in signal recovery.

Here $r(t)$ is the reference signal, $s(t)$ the signal of interest and $n(t)$ represents the effect of additive noise and interference. The high-order products of multiplication are suppressed by a low-pass filter in the system output.

When $r(t)$ and $s(t)$ are closely correlated the term $r(t)s(t)$ gives rise to a distinctive response. For example, if the characteristics of the signal are sensibly fixed during some stage of an experiment the final output will be a constant voltage which is proportional to signal level.

Ideally, there is no correlation between $r(t)$ and $n(t)$. In this case the average value of the noise product is always zero in the final output. The response to a fixed signal is therefore quite unambiguous because there is no error due to rectified noise components: any residual fluctuations due to noise appear as an AC variation which does not affect the mean value of the 'true' output due to the signal.

In principle, these residual fluctuations can always be attenuated to an acceptable level by reducing the bandwidth of the output low-pass filter. In lock-in systems this represents the major source of signal-to-noise improvement. Other demodulation techniques make no fundamental distinction between signals and noise and so must be provided with noise suppression filters to ensure a high signal-to-noise ratio *before* detection. In difficult conditions this may imply using highly selective filters centred on the signal frequency which render rectifier systems incapable of operation if the signal frequency is changing by accident or design. In contrast, lock-in systems can maintain a very small noise rejection bandwidth even when the signal is changing over many decades of frequency. Provided that a reference signal can be generated, lock-in systems are ideally suited to the role of 'tracking' signals in noise.

3. Phase-sensitive detection

In practical systems the multiplication function which is essential to the principle of synchronous detection is provided by a high-quality mixer or *phase-sensitive detector*. The phase-sensitive detector gives selective rectification of the signal by means of a switching operation as shown in figure 2.

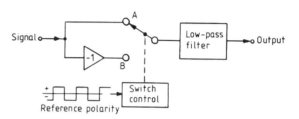

Figure 2. Block diagram of a switching phase-sensitive detector.

The two-position switch is controlled electronically from the reference voltage and switches between A and B as the reference waveform changes polarity. This gives a systematic change of gain in the signal path between $+1$ and -1. When the signal and reference are derived from the same source and thus coherent the switching operation gives rise to the familiar voltages summarised in figure 3.

A switching multiplier is essential in systems which are expected to maintain high accuracy and linearity under conditions of adverse signal-to-noise ratio. The main purpose of a *reference channel* is thus to respond to the zero crossings of

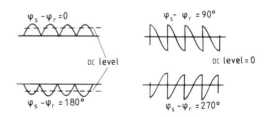

Figure 3. Voltage waveforms appearing at the output of a phase-sensitive detector for a sinusoidal signal and a coherent reference. The response depends upon signal magnitude and the relative phase, $\varphi_S - \varphi_R$, of the signal and reference inputs.

the reference input voltage and to generate a precise square-wave switching voltage. A reference phase shifter is provided to control the relative phase of the two inputs to the phase-sensitive detector. The response of this type of system is generally well known; for a coherent sine wave signal of amplitude V_s at relative phase θ, the low-pass filter transmits a constant output voltage:

$$v_0 \propto V_s \cos \theta.$$

The response is therefore a maximum when the signal and coherent reference are brought in-phase at the phase-sensitive detector input.

The response of the phase-sensitive detector to noise and other asynchronous signals have been dealt with by Blair and Sydenham (1975). In general it is only those spurious components which lie in the frequency 'windows' adjacent to the reference frequency and its odd harmonics which can perturb the final output.

The transmission windows for a switching phase-sensitive detector have the response magnitudes shown in figure 4. The responses are proportional to the amplitude of the associated reference Fourier component and the shape is determined by the frequency response, $H(j\omega)$ of the low-pass filter. Figure 5 shows the form of the windows when the low-pass filter consists of a simple RC section with time constant T and radian cut-off frequency $1/T$.

An unwanted component falling within a transmission

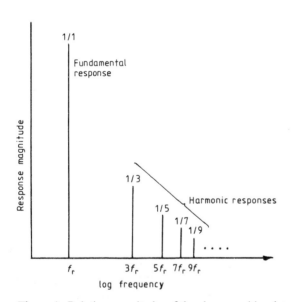

Figure 4. Relative magnitudes of the phase-sensitive detector transmission 'windows' for a conventional 'squarewave' reference.

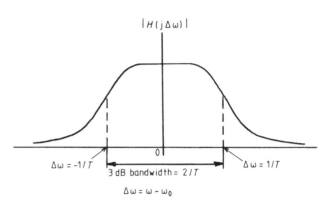

Figure 5. Frequency response of a single transmission window when the low-pass filter is a simple RC filter with $T = RC$. $\omega_0 = (2n + 1)\omega_r$.

window gives rise to an alternating 'beat' note in the final output. In order to produce a 'true' DC response however a signal must be coherent with one or more of the reference Fourier components.

For recovery in *white noise* the *noise bandwidth* of the low-pass filter, B_N, becomes important. If a signal is accompanied by white noise with input noise bandwidth B_I, the *improvement* in signal-to-noise ratio brought about by synchronous detection is then, simply, B_I/B_N. $B_N = 1/4T$ for a single-section RC filter and $1/8T$ for a two-stage filter. Since T is commonly 1 s or greater in typical recovery applications we find that enormous improvement factors can be obtained without the need for 'tuning' or other critical adjustments.

4. Two-phase systems

Two-phase lock-in amplifiers incorporate a pair of phase-sensitive detectors operated in quadrature as shown in figure 6. This arrangement enables both the *in-phase* and *quadrature* components V_A and V_B of a coherent signal to be measured simultaneously.

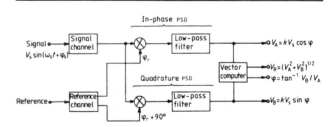

Figure 6. A two-phase lock-in system. The reference channel incorporates all facilities for squarewave generation and phase-shifting. k = scaling factor; $\varphi = \varphi_s - \varphi_r$.

Usually, provision is made for a *vector computer* which calculates the resultant:

$$V_{\mathrm{R}} = \sqrt{V_{\mathrm{A}}^2 + V_{\mathrm{B}}^2}$$

and the phase

$$\phi = \tan^{-1} V_{\mathrm{B}}/V_{\mathrm{A}}.$$

Two-phase systems can thus display their output in either cartesian or polar form. The polar form is important from an operational point of view because the amplitude of a coherent signal can be measured continuously without the need for phase adjustment.

When using a vector computer, asynchronous signals generate quadrature 'beat' components within the output low-pass filters which give rise to a DC output. The lock-in can therefore function as a *wave analyser* and provision of a swept reference makes wide-band spectrum analysis possible.

In signal recovery applications this DC response to asynchronous signals can be most undesirable. Also in conventional systems, the vector computer is applicable only when the signal has sine wave form. In view of this many two-phase systems incorporate the *vector tracking* arrangement shown in figure 7.

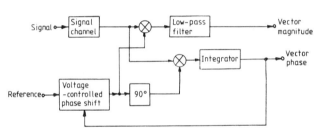

Figure 7. Configuration of a two-phase lock-in system to give 'vector tracking'.

Vector tracking can be selected by push-button in many two-phase systems. The quadrature phase-sensitive detector is used to provide a control signal to a voltage-controlled phase-shifter and the loop is arranged to keep signal and reference in phase at the other phase-sensitive detector. The control signal provides an output voltage proportional to relative phase and the in-phase output is proportional to the signal magnitude, free of errors due to rectified noise. A control range of $\pm 100°$ is possible in most systems without need for adjustment.

Modern two-phase systems offer the possibility of precision measurement over many decades of frequency. This capability has opened up several new application areas, notably for signal analysis and transfer function analysis.

5. 'Conventional' lock-in systems: key specifications

5.1. Introduction

The main advances in lock-in instrumentation have been achieved for phase-sensitive detectors operating in the frequency range up to 1 MHz and most commercial systems are restricted to an upper frequency limit of a few hundred kHz. It has become common practice for manufacturers to quote specification at 'mid-band', usually at 1 kHz, and this will apply to many of the 'typical' figures given here.

5.2. Dynamic reserve

If a phase-sensitive detector is not highly linear then the presence of asynchronous signals will cause DC shifts at the output which are indistinguishable from the 'true' response due to a coherent signal.

In practice, modern phase-sensitive detectors display excellent linearity within a specified range of signal and noise voltages and we identify a 'maximum allowed' level of asynchronous signal. This is the level which causes an error of 5% in the response due to a full-scale coherent input.

As might be expected this maximum allowed asynchronous level is far greater than the level of coherent signal required for a full-scale output. A wide input range is thus held 'in reserve' for noise and interference components, an idea which leads to the definition of dynamic reserve:

$$D_{\mathrm{R}} = \frac{\text{max. allowed peak–peak level of asynchronous signal}}{\text{peak–peak level of coherent signal giving full-scale output}}.$$

Dynamic reserves of 60 to 80 dB ($\times 1000$–10 000) are typical of modern phase-sensitive detectors. The 'maximum allowed' level is almost invariably measured by using an interfering sine wave and the specification is presented by plotting the measured dynamic reserve as a function of interference frequency as shown in figure 8. For a broad-band phase-sensitive detector

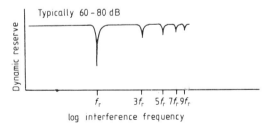

Figure 8. Dynamic reserve of a broad-band phase-sensitive detector shown as a function of interference frequency. The system is most susceptible to interference close to the reference frequency and its odd harmonics.

with a conventional switching action the resulting curve is flat over a wide frequency range with local 'dips' in the vicinity of the transmission windows centred on odd harmonics of the reference frequency.

5.3. Output stability

If the total signal input to the phase-sensitive detector is removed and the reference waveform left connected the output should, ideally, fall to zero. The application of a millivoltmeter to the output will, however, reveal a residual offset voltage and long-term observation will show the effect of drift in the output. The offset is, moreover, dependent upon the reference frequency and will be found to increase quite markedly above 10 kHz, an effect due to capacitive coupling of switching spikes from the reference input through to the final output.

Provision is usually made so that the offset voltage can be trimmed to zero under a particular set of operating conditions, leaving the drift component as the limiting factor to precision. The drift is usually labelled as output 'stability' and quoted as a fraction of the full-scale output voltage per degree Kelvin.

If we now consider the problem of measuring either very small coherent signals or small *changes* in a coherent signal, the measurement uncertainty is absolute when the corresponding output change is comparable with the drift specification of the phase-sensitive detector. The *minimum detectable* signal is then defined as

$$S_{min} = \text{output stability} \times \text{(coherent signal to give full-scale output)}.$$

5.4. System considerations

In order to achieve maximum flexibility the phase-sensitive detector is almost invariably operated with a signal channel amplifier having switched gain and provision for signal conditioning filters. In addition, the gain of the phase-sensitive detector may also be variable, by the inclusion of a variable-gain DC amplifier *after* the phase-sensitive detector. The DC amplifier serves to 'expand' the output and is referred to as an *expand amplifier*. These additions lead us to the basic 'conventional' lock-in amplifier shown in figure 9.

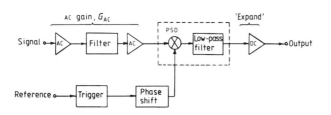

Figure 9. A basic lock-in amplifier incorporating signal channel filters and an output DC 'expand' amplifier.

A typical system might offer up to twelve switched range positions in the signal channel AC amplifier and perhaps three different levels of output expansion. It follows therefore that there must be a number of possible combinations of AC and DC gain which give the same overall system sensitivity and that the signal recovery capability of the system must be influenced accordingly.

For example, if the signal at the amplifier input is accompanied by a high level of noise, then the use of high signal channel gain, G_{AC}, may result in premature overload at the phase-sensitive detector input. Conversely, the adverse effect of noise at the phase-sensitive detector input can often be reduced by using a smaller value of AC gain. The reduced response to a coherent signal can then be 'expanded' using the DC output amplifier in order to restore the overall sensitivity of the system. For a fixed level of coherent input signal *to the amplifier* we find that the system can tolerate progressively higher input noise levels as G_{AC} is reduced and G_{DC} is increased to maintain constant overall gain. In this way, the dynamic reserve measured at the amplifier input can be controlled via the gain selection. For ease of operation most commercial systems incorporate a calibrated dynamic reserve switch whereby the gain distribution can be altered without affecting the overall system sensitivity.

5.5. Dynamic reserve/output stability trade-off

The output stability of a phase-sensitive detector system is degraded when the DC output gain is increased so as to improve the dynamic reserve. One of the prime objectives of phase-sensitive detector systems design is therefore to achieve high dynamic reserve linked with good output stability. For a basic broad-band system a combination of 60 dB dynamic reserve and 10 PPM K^{-1} output stability is generally considered to be 'good' practice. When comparing competing systems, the ratio of these two specifications provides a useful figure of merit.

The ability to control dynamic reserve and output stability by changing the system gain configuration has important operational consequences. For example, when signals are relatively 'clean', dynamic reserve is not at a premium and can be traded for good output stability for precision measurements. In contrast, noisy signals may require a higher dynamic reserve capability, but here output stability may not prove to be a limiting factor. It should be noted that dynamic reserves of 60 dB and greater are usually aimed at signals perturbed by *discrete* interference components. The recovery of signals from equivalent levels of broad-band noise is theoretically possible but requires the use of extremely narrow-band output filters and a correspondingly long observation time.

5.6. Signal conditioning and overload capability

Signal conditioning refers to the elimination of out-of-band noise components by selective filtering. This is often an essential step if a large amplification factor is required to bring the signal of interest to a level compatible with phase-sensitive detection.

By eliminating noise components *before* detection the dynamic reserve of the system can appear to be much greater than in 'broad-band' operation. We now refer to the *overload* capability of the system which should always be specified with respect to a particular set of operating conditions. The maximum overload capability is determined by the maximum peak–peak value of asynchronous signal which the amplifier can withstand at its input, relative to a full-scale coherent signal at the same point.

5.7. Dynamic range

It is usual to define input and output dynamic range for a lock-in amplifier. The relationship between these specifications, dynamic reserve and minimum detectable signal is clarified in figure 10.

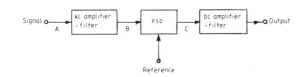

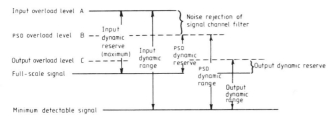

Figure 10. Fundamental relationships between dynamic reserve, minimum detectable signal and dynamic range in a lock-in recovery system.

5.8. Reference channel specifications

Key specification points are low *phase noise* to eliminate undesirable jitter at the point of detection and low *phase drift* to enable precision measurements to be made over a period of time.

In broad-band systems the relative phase of signal and reference can be maintained to less than 1 degree over many decades of frequency. In two-phase systems the quadrature accuracy of the two channels is of extreme importance and is normally better than 0.1 degree.

The ability of the reference channel to maintain good phase accuracy while tracking a signal of changing frequency is expressed in terms of reference channel *slew rate*. Here, the maximum allowable rate of change of signal frequency is quoted, usually for a phase 'slip' of around 5 degrees.

6. Modern lock-in amplifier configurations

6.1. Introduction

In the last decade there has been a growing awareness of the

shortcomings of the conventional lock-in systems described so far. For example, the dynamic reserve is often insufficient for signals perturbed by massive interference components. Also, operation at maximum reserve implies that the output stability is degraded and it is often difficult to exploit the full overload capability of the system by using signal conditioning filters.

The harmonic responses can limit system capability in many applications such as the infrared detection system shown in figure 11. Here the phase-sensitive detector is susceptible to spurious components at mains frequency which coincide with the fifth harmonic of the excitation frequency. At such low frequencies there is little scope for improvement owing to the narrow spacing of the transmission windows.

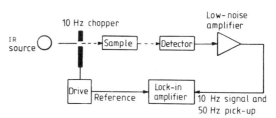

Figure 11. A low-frequency optical experiment. A conventional lock-in amplifier will be susceptible to interference at mains frequency when the reference is close to 10 Hz.

Other applications to suffer from multiple responses are signal analysis and AC bridge measurements. In the latter case a phase-sensitive detector responds to harmonic components of the excitation oscillator even when the fundamental component is satisfactorily nulled.

Other specification areas can be found wanting in specific applications. Thus in swept-frequency analyses the limited slew rate of the reference channel can be a major factor. There are, moreover, several areas which could benefit from lock-in technology but where a reference signal is not readily available, for example, in communications systems and other systems where the detector is remote from the excitation source.

Recognition of these factors has resulted in the emergence of several 'advanced' systems which aim to improve performance in one or more critical specification areas. The broad relationships between the different systems is illustrated in figure 12. In the main, we find that attention is given to the processing of the signal and reference waveforms, leaving the basic, switching, phase-sensitive detector intact at the heart of the system.

6.2. 'No reference' systems

In cases where a signal is accompanied by less than about 30 dB of broad-band noise it is feasible to generate a reference locally using a phase-lock loop as shown in figure 13.

The purpose of the loop is to pull a voltage-controlled oscillator (vco) into synchronisation with the signal. The dynamics of the loop and the rejection of noise are controlled by the characteristics of the loop filter and the action of the loop is to bring the vco into quadrature with the signal in the locked condition.

Some lock-in amplifiers incorporate a vco and phase-detector in the reference channel. The loop filter is not generally accessible but an adjustment is provided to trim the vco centre frequency and so aid locking.

In other lock-in amplifiers the phase-sensitive detector can be used as a building block in a phase-lock system using an external vco. Push-button selection converts the output filter to a lag–lead loop filter controlled by the time-constant switch.

In a two-phase system one phase-sensitive detector is often made available for phase-locking leaving the other phase-sensitive detector free to measure the amplitude of the locking signal.

6.3. Heterodyne systems

6.3.1. Introduction. The overload capability of a lock-in system can always be improved by using a bandpass filter centred on the signal frequency. Since the signal is effectively 'stripped' of all harmonic components this filter is also effective in suppressing harmonic responses even when the phase-sensitive detector has a switching reference waveform. The resulting system is said to have 'fundamental-only' response.

Such a filter is difficult to set up and tune and its inclusion

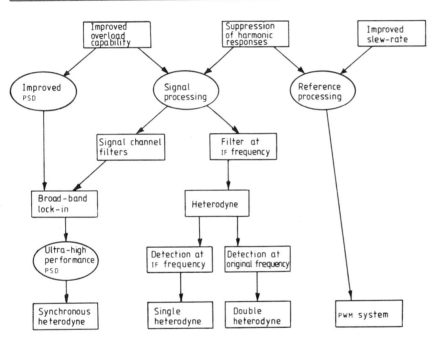

Figure 12. Classification chart for modern lock-in systems.

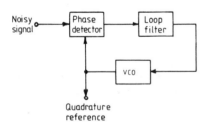

Figure 13. A basic phase-lock loop.

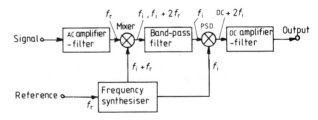

Figure 14. A single heterodyne lock-in amplifier.

can lead to serious errors due to phase-shifts even for small signal frequency variations. In order to obtain the benefits of fundamental-only response, and to regain the ability to track signals of changing frequency, several systems have been developed to operate on the heterodyne principle. Here, a mixer is used to translate the input signal up to a relatively high frequency, the intermediate frequency. In figure 12 heterodyne systems were broadly classified according to whether the phase-sensitive detector operates at the IF or at the original signal frequency. These alternative approaches can be implemented using 'single' and 'double' heterodyne systems.

6.3.2. Single heterodyne systems. Here, a single conversion is used to bring the signal to the intermediate frequency as shown in figure 14. System complexity is increased largely on acount of the need to implement a local oscillator of precise frequency $f_i + f_r$. The IF filter rejects unwanted components in the mixer output prior to phase-sensitive detection. Reference phase-shifting can be introduced at the original reference frequency, at the output of the synthesiser – at $f_i + f_r$ – or at the intermediate frequency. The advantage of the latter is that the phase-shift circuits operate at a fixed frequency and the need for wide-band operation is avoided. The output stage consists of a conventional low-pass filter (time-constant control) and DC amplifier.

Mixer products at $f_i + 2f_r$ will be transmitted by the IF filter when f_r is sufficiently low but have no effect on the DC response of the phase-sensitive detector. More serious is the effect of a spurious input at frequency $2f_i + f_r$ which generates an 'image' component at the output of the signal mixer. For this reason heterodyne systems are operated with a sharp cut-off 'image reject' filter in the signal channel, giving severe attenuation for inputs of frequency greater than about $f_i/5$. The image reject filter contributes to the overload capability of heterodyne systems which is generally very high for interference signals above the maximum reference frequency. In contrast, the IF filter has minimal effect on overload capability over much of the operating frequency range. The IF filter characteristics are selected in the interest of suppressing spurious responses associated with the mixer stage which is invariably of the switching-mixer variety.

6.3.3. Double heterodyne systems. The single heterodyne system achieves fundamental-only response because the intermediate frequency bears no harmonic relationship to the frequency of the signal and reference.

In the double heterodyne system shown in figure 15, the final detection takes place at the signal frequency. Harmonic responses are rejected in this case because of the 'stripping' effect of the IF filter referred to earlier.

In audio-frequency lock-ins the intermediate frequency of the double heterodyne is typically 1 MHz and a crystal filter is used to achieve very small bandwidths. The bandwidth is usually of the order of 10 Hz so that harmonic rejection is not so effective at low signal frequencies.

The use of heterodyne schemes can increase system overload capability to as high as 120 dB and give harmonic rejection in excess of 60 dB. In some systems, however, this improved performance is obtained at the expense of phase accuracy and increased phase jitter compared with 'conventional' broad-band systems.

6.4. PWM systems

The harmonic responses of a switching phase-sensitive detector result from the abrupt change of gain as the reference waveform changes polarity. In practice the effect of a *sinusoidal* variation in gain can be obtained by using a high-frequency switching waveform which is *pulse-width modulated* by a sinewave voltage. The result is a system with fundamental-only response which makes use of a standard phase-sensitive detector and an unmodified signal channel. The system will be free from spurious responses provided that the reference switching frequency is much higher than the anticipated range of signal and interference frequencies.

The essentials of a pulse-width modulated system are shown in figure 16. The reference channel consists of the usual trigger circuit, squarewave generator and phase-shifter up to the sine wave synthesiser which generates a low-frequency sine wave precisely synchronised to the phase-shifter output. Carter and Faulkner (1977b) have described circuits to fulfil this task.

Advantages of the PWM approach are its relative cheapness

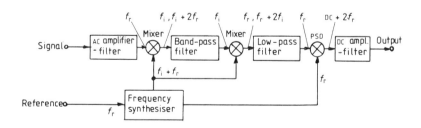

Figure 15. A double heterodyne lock-in amplifier. Detection takes place at the original signal frequency.

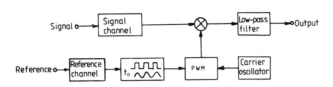

Figure 16. Essential features of a lock-in amplifier operating with a pulse-width-modulated reference.

and that it can be included as an option in an otherwise conventional lock-in amplifier. An attractive feature is that the pulse-width modulator can be driven directly by an external source, giving the ability to achieve correlation-detection with non-sinusoidal signals. Also, since the reference phase-shift circuits are now bypassed, the major source of slew-rate limitation is eliminated and it becomes possible to obtain fundamental-only response with very fast sweep waveforms. Two-phase lock-ins with PWM facility have thereby found new application areas for wide-band spectrum analysis.

6.5. Synchronous heterodyning

Synchronous heterodyning is an approach which improves the trade-off between dynamic reserve and output stability which is inevitable in conventional systems. A system which has been used commercially is shown in figure 17.

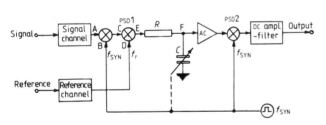

Figure 17. Synchronous heterodyning.

The effect of the first mixer is to alternately reverse the phase of the incoming signal at a frequency f_{SYN} (normally much less than f_r). The output of phase-sensitive detector 1 is now the AC signal shown in figure 18(e). The 'rotating capacitor' filter is synchronised to f_{SYN} and so transmits the square-wave component shown in figure 18(f) with an amplitude proportional to the signal of interest. This can be amplified in an AC amplifier without incurring a drift penalty before final detection takes place in phase-sensitive detector 2.

Using this technique it has been possible to obtain dynamic reserves of $\times 300\,000$ consistent with output stabilities of 1000 PPM K^{-1}. This is equivalent to operation with an overall dynamic range (§ 5.7) approaching 170 dB. It is popularly believed that synchronous heterodyne systems are inherently free from harmonic responses but this is clearly not the case. Indeed, the system described was marred by a series of spurious responses at certain critical combinations of f_{SYN} and f_r and their harmonics.

Interest in synchronous heterodyning has recently been revived as a means of improving the dynamic range of phase-sensitive detectors operating at high frequencies. The technique has been applied to phase-sensitive detectors operating at intermediate frequencies up to 1 MHz in commercial heterodyne lock-in amplifiers.

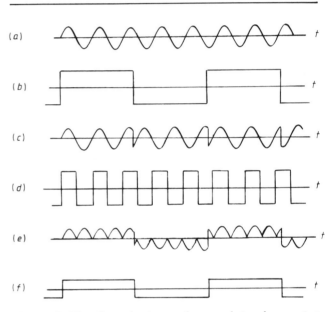

Figure 18. Waveforms in the synchronous heterodyne system of figure 17.

7. Computer-controlled systems

7.1. Introduction

The internal arrangement of commercial lock-in systems has long been suited to some form of remote control. This is because all major switching functions in the signal and reference channels have been provided in the form of FET switches and the reference channel phase-shifter has usually been designed for voltage control.

For several years instruments have been available with 'program' options whereby the front panel controls could be duplicated by applying logic levels to a rear panel connector.

Instruments fitted with an analogue-to-digital converter (ADC), digital display and interface to peripheral printers have also been a standard catalogue feature. The advent of the microprocessor has now resulted in a new generation of instruments in which the switching functions and the normal user routines can be provided from a preprogrammed microprocessor unit. In addition to the standard controls the user then has the option to switch to automatic operation where the lock-in system automatically switches sensitivity and phase in response to an applied signal.

An alternative mode is to use a microprocessor or some other logic unit to interpret commands from the IEEE or IEC bus, bringing the lock-in system under the control of a central computer-controller. This has opened up opportunities in the automatic test environment and has vastly increased the flexibility of lock-in systems for use in complex installations for routine analysis.

7.2. Automatic routines

In automatic operation the programmed routines of principal interest are those for *autoranging* (adjustment of sensitivity), *autophasing* (adjustment of phase) and *autoset*, a combination of the first two. To achieve autoranging of a noisy signal, all switching decisions are made from readings of the final system output. The programs therefore follow closely the routines which would be performed in normal, manual, operation from the front panel. In signal recovery applications, the need for narrow output bandwidth and correspondingly long time constants greatly increases the response time of the lock-in system. The programs take account of the time constant setting; as a result, automatic adjustment of the system can take a time which is comparable with manual control.

Even under favourable conditions the output from a lock-in might be perturbed by residual noise. For this reason it is necessary to introduce *hysteresis* bands at the decision levels for range switching. Failure to do this could result in highly unstable operation with noisy signals. In practice the hysteresis bands can be set manually or from the IEEE/IEC bus and the sensitivity switching can be programmed to give either best output stability (high AC gain) or best dynamic reserve (high DC gain). Full information about the state of the instrument including sensitivity, time constant output voltage, phase, etc., can be transferred to peripheral equipment.

7.3. Examples of lock-in amplifiers in computer-controlled systems

7.3.1. Photoacoustic spectroscopy. The long-standing association between lock-in amplifiers and the general field of spectroscopy has been extended in recent years with the increasing interest in photoacoustic spectroscopy.

The photoacoustic effect involves the generation of an acoustic signal by a sample exposed to modulated light. By providing a suitable sample chamber the resulting pressure modulation can be detected by means of a sensitive microphone and measured using a lock-in amplifier. The technique has proved particularly valuable for evaluating opaque samples and samples which scatter light such as powders and turbulent liquids (Rosencwaig 1975).

Munroe and Reichard (1977) have described a photoacoustic spectrometer which has formed the basis of a commercial 'turn key' system. The spectrometer incorporates two two-phase lock-in amplifiers with fundamental-only response. One is used for measuring the microphone signal and the other measures the output from a pyroelectric detector exposed to the xenon lamp source. The outputs from the lock-ins and from other analogue sources are multiplexed to a 14-bit ADC and all operations are under microprocessor control with keyboard entry of experimental parameters. As with all systems of this type the object is to give fully automatic spectroscopy with the final data properly compensated, normalised and calibrated.

7.3.2. 'Batch' testing of optical detectors. A recent industrial application has involved the use of lock-in amplifiers in testing batches of optical detectors. The lock-in provided the means of measuring sensitivity relative to a standard light source and was also used in a noise measurement mode to evaluate signal-to-noise ratios at specified operating frequencies. The equipment had a 'manual' operating mode and could be linked, together with other similar systems, to a central computer-controller for storage and comparison of data and to test for ACCEPT/REJECT conditions.

7.3.3. Fibre-optic test equipment. Lock-in techniques are ideally suited to the measurement of attenuation in lengths of optical fibre, using 'chopped' radiation from an optical source and a silicon or germanium detector. Commercial systems are now available which incorporate a microprocessor to control wavelength selection (by monochromator or prefilter), detector position — for measurement of numerical aperture — and to automate the selection of sensitivity in the lock-in system. These systems incorporate the optics and fibre manipulators necessary for the routine measurement of samples in an industrial environment. System capability can be extended by using an interface to a control computer-controller.

8. Conclusions

Modern lock-in amplifiers offer several operational advantages over conventional systems, notably in the areas of overload capability and suppression of harmonic responses. Despite this, conventional systems are often superior in terms of phase precision and true wide-band operation and represent an economic solution to many measurement problems.

Acknowledgments

The author wishes to acknowledge the cooperation of Simon

Carter, now with Hewlett Packard Limited, South Queensferry, in planning this review.

References

Blair D P and Sydenham P H 1975
J. Phys. E: Sci. Instrum. **8** 621–7

Carter S F and Faulkner E A 1977a
Electron. Lett. **13** 339–40

Carter S F and Faulkner E A 1977b
Electron. Lett. **13** 381–2

Fisher E H 1977
Laser Focus **11** 82–8

Hieftje G M 1972a
Anal. Chem. **44** (6) 81–8

Hieftje G M 1972b
Anal. Chem. **44** (7) 69–78

Munroe D M 1973 The heterodyning lock-in analyser
Ithaco Corp. Bulletin

Munroe D M and Reichard H S 1977
American Laboratory **2** 119–31

Rosencwaig A 1975
Physics Today **28** (9) 23–30

Taub H and Schilling D L 1971 *Principles of Communication Systems* (New York: McGraw-Hill)

Chapter 8

Correlation in instruments: cross correlation flowmeters

M S Beck

Department of Instrumentation and Analytical Science,
University of Manchester Institute of Science and
Technology, Manchester M60 1QD, England

Abstract The basic principles are briefly explained,
showing how cross correlation can be used to identify
dynamic characteristics of a wide range of telecommunication,
structural and process systems. The main emphasis of the
paper is on cross correlation flowmeters, which are
developing to a stage where they can successfully solve
industrial and environmental measurement problems,
ranging from the flow of highly polluted liquids in pipes to
the flow of gas from volcanic jets.

Cross correlation flowmeters are based on measuring the
transit time of a tagging signal (turbulence, clumps of
particles, etc.) in the flow between two axially separated
sensors. The transit time is measured by a cross correlator.
The design of the various subsystems is discussed in some
detail and the various sensing techniques required for
specific applications are described. Cross correlation
flowmeters have become a realistic proposition because of
the reducing cost of large scale integrated circuits and
microprocessors from which the correlator may be realised,
a number of different designs of cross correlator are
described and their relative merits discussed.

1 Introduction

Statisticians have used cross correlation to estimate the cause
and effect relations in sets of population data (e.g. the dependence of crop yield on previous rainfall). Engineering applications of cross correlation have increased during the past few
decades, no doubt because of the availability of inexpensive
computers to produce cross correlation results in 'real time'
(e.g. for recovery of information in weak radar echoes,
identifying the dynamics of chemical processes, etc.). Lange
(1967) outlines a wide range of cross correlation applications.

The basic concept is simple. The dynamic relationship is the
way in which values of the input signal affect future values of
the output signal. The response to a unit (dirac) impulse is the
simplest embodiment of this, unfortunately many systems
cannot have unit impulses applied which give a clearly dis-
cernible output response (Eykhoff (1974) presents an excellent
account of system identification problems). By regarding the
normal (randomly time varying) input as a series of impulses,
merging to give the continuous input signal $x(t)$ and convoluting (i.e. multiplying and integrating with suitable time shifts)
this with the system output $y(t)$, we obtain the cross correlation function R_{xy} which, assuming the input signal has a wide
bandwidth, contains all the information about the dynamics
of the system. The equation for this cross correlation is:

$$\text{cross correlation } R_{xy}(\tau) = \frac{1}{T} \int_0^T x(t-\tau)y(t) \, dt.$$

A specific advantage of cross correlation is that any spurious
signals present as interference on the measured output $y(t)$
are strongly rejected because they have no correlation with the
input signal $x(t)$. The process by which the spurious signals
are rejected will now be considered in more detail. Let the
measured output $y(t)$ be equal to the sum of a (desired) noise
free signal $y'(t)$ and an (undesired but unavoidable) spurious
noise interference $z(t)$. i.e.

$$y(t) = y'(t) + z(t).$$

Substituting in the cross correlation equation gives:

$$R_{xy}(\tau) = \frac{1}{T} \int_0^T x(t-\tau)[y'(t) + z(t)] \, dt.$$

This can be split into two integrations hence:

$$R_{xy}(\tau) = \frac{1}{T} \int_0^T x(t-\tau)y'(t)\,dt + \frac{1}{T} \int_0^T x(t-\tau)z(t)\,dt.$$

Now the second integral represents the cross correlation between the system input $x(t)$ and a spurious signal $z(t)$. Since these are not related their cross correlation tends to zero provided that the integration time T is sufficiently long. This rejection of spurious interference is the most important advantage of cross correlation.

The cross correlation method of determining system dynamics has many similarities to the frequency response method. The former determines the relationship for specific time delays, whereas the latter determines the relationship for specific frequencies. The cross correlation and the frequency response functions are related by 'Fourier type' transforms, so provided one function is known, the other can be readily computed. The reader should refer to the book by Bendat and Piersol (1971) for a complete treatment of these procedures for random data analysis.

In practice, direct calculation of the frequency response has been used mainly for structural vibration analysis, control system design by Bode, Nyquist diagrams, etc. (Shinners 1973) and telecommunication channel response measurement. Cross correlation is more directly used where measurements requiring some knowledge of time delays are required, such as velocity measurement of steel (Butterfield *et al* 1961), paper strip (Kashiwagi 1968), road and rail vehicle speed (Zimmer *et al* 1976, Idowaga and Ono 1976), range measurement by sonar and radar (Arthur *et al* 1979). This paper will describe one of the major uses of cross correlation, its application to flow measurement, and in so doing will illustrate the advantages and problems associated with the use of correlation for measurement purposes. General design principles, methods of implementation, means of on-line computation, and accuracy of measurement will be discussed.

1.1 *Cross correlation for flow measurement of difficult fluids*

The process industries handle a wide range of fluids from clean liquids and gases to aggressive and difficult fluids such as hydraulically and pneumatically conveyed solid materials, sewage etc. Measurement of flow and other process variables in industries handling clean fluids has been comparatively easy and this has led to a great deal of automatic control in such processes. On the other hand industries handling solid materials and difficult fluids have found that process measurement has been very difficult and this has hampered the use of automatic control in these industries. It seems certain that the main effort in process automation during the next decade will be in industries handling solid materials. This trend will be accentuated by an increasing use of solid fuels as our liquid and gas fuels become depleted.

Essential features of flowmeters for measuring flows of difficult fluids in pipes are that they should not obstruct the flow, which would give a risk of blockage and they should not have any parts which wear out by being in abrasive contact with the fluid (Beck *et al* 1978). In other cases the measuring instrument must even be truly remote from the fluid, consider for example the measurement of the flow of hot gases from steel making furnaces (Webb *et al* 1976).

Referring to figure 1 cross correlation flowmeters measure flow by sensing the passage of some disturbance or tagging signal in the flow at one position, and measuring how long it takes that disturbance to travel to a downstream point. This is basically a very simple principle and fortunately the tagging signals can be detected by sensors which do not contact the fluid.

The cross correlation function as defined above is:

$$R_{xy}(\tau) = \frac{1}{T} \int_0^T x(t-\tau)y(t)\,dt.$$

It can be shown that this function has a maximum value when the cross correlation lag τ is equal to the transit time τ^* of the tagging signals, hence the flow velocity is given by

$$u = l/\tau^*$$

where l is the spacing of the sensors A and B.

A proof that the time delay of the cross correlation peak is equal to the fluid transit time τ^* is given in an early publication by Beck *et al* (1968).

The cross correlation calculation has in the past required quite elaborate and expensive computing systems; now, fortunately, modern large scale integrated circuits enable cheap cross correlators to be made. The design of cross correlators will be reviewed in §4. Before this we will consider alternative flowmeters (§1.2), total system design (§2), and the selection of suitable sensors (§3).

1.2 *Alternatives to cross correlation flowmeters for difficult fluids*

The flow of a fluid is often measured by putting some obstruction into the flow (Open University T291). Orifice plates, venturis and vortex flowmeters all fall into this category. They are often unreliable due to wear by abrasion when the fluid contains suspended solid particles; they can also cause solids to drop out, leading to a complete stoppage of flow. The pressure loss caused by the obstruction leads to an increase in the energy required to pump the fluid, this can be significant with large flows of water, sewage, natural gas etc. In contrast there is no pressure loss with cross correlation flowmeters.

Ultrasonic pulse transit flowmeters (Cousins 1978) seem to be promising for difficult fluids. The generally preferred arrangement uses piezoelectric transducers located diagonally across the pipe which transmit ultrasound pulses upstream against the flow and downstream with the flow. The difference between the upstream and downstream pulse transit times is a measure of the flow. However, because the velocity of sound

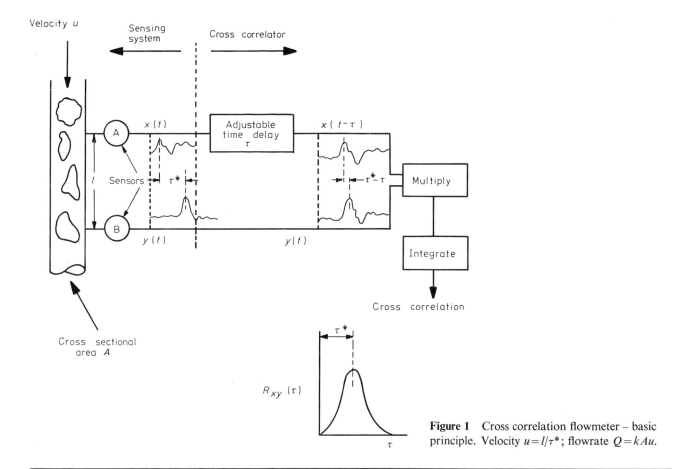

Figure 1 Cross correlation flowmeter – basic principle. Velocity $u = l/\tau^*$; flowrate $Q = kAu$.

is much greater than the velocity of the fluid these devices involve measuring a small difference between two large quantities which has in the past been a cause of considerable error, although it is claimed that newer systems have reduced this error (Watts 1980). In order to obtain an accurate transmission path the piezoelectric transducers have to be in contact with the fluid, in contrast with ultrasonic cross correlation flowmeters in which the transducers can be clamped onto the outside of the pipe.

The electromagnetic flowmeter does not obstruct the flow and is possibly the only commercially available instrument which has been proved successful in measuring the flow of difficult fluids (De Jong 1978). However, it can only be used with electrically conducting fluids, and it is very expensive for the larger pipe sizes. In contrast, with suitable sensor design, cross correlation flowmeters can measure virtually any fluid and their cost does not increase significantly as the size of the pipe increases. In addition, cross correlation flowmeters rely on the fundamental measurements of time and distance and do not require calibrating for individual applications, this

contrasts strongly with electromagnetic flowmeters which have to be individually calibrated for the specific pipe size in use and which may require recalibration when a fault is suspected.

2 Design of cross correlation flow meters

Our initial development of cross correlation flow meters was done on a largely empirical basis and a fuller theoretical framework was built at a later stage. In my opinion, the quickest way to develop a new technological area – such as cross correlation flowmeters – is to develop experimentally instruments for a range of applications. After the early development phase, the sight of a few experimental prototype instruments actually working helps to stimulate the more theoretical approach which is required to design the optimum cost-effective systems.

In the following section we will model the whole measurement system and break it down into subsystems in order to illustrate the more fundamental aspects of designing cross correlation flowmeters. Further details of modelling correla-

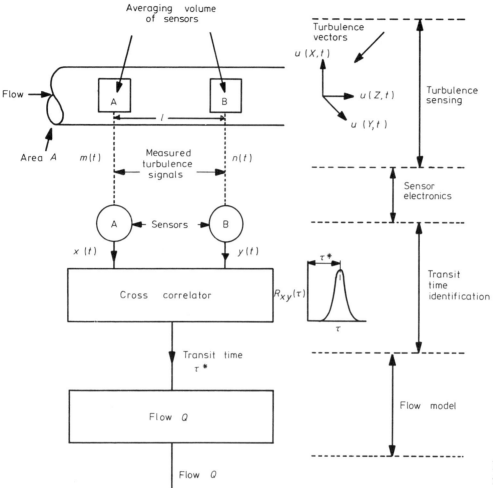

Figure 2 Cross correlation flowmeter – schematic of subsystems.

tion flow measurement systems are given by Fischer and Davies (1964) and Kippham and Mesch (1978).

2.1 *Model of the measurement system*

Referring to figure 2, a cross correlation flowmeter measures the flow velocity by obtaining the transit time of a tagging signal between sensors at locations A and B. The following flow model is used to derive the measured flow Q from the measured transit time τ^*—

$$Q = k(l/\tau^*)A \qquad (1)$$

where l = spacing of sensors, A = cross sectional area of pipe, k = a calibration factor, mainly to allow for non-uniformity of the velocity profile (see §5.1).

The tagging signal is generated by the fluid turbulence, the turbulent eddies can be detected, for example, by the modula-

tion they superimpose (by doppler effect) on a beam of ultrasound perpendicular to the direction of flow. Alternatively particles can become entrained in the turbulent eddies to form clouds (just as clouds are formed in the atmosphere by the effects of atmospheric turbulence on the water droplets); these clouds can form the tagging signal and they may be detected by ultrasonic or optical beams, electrical conductivity, electrical capacitance effects etc. The design of sensors for these signals is discussed in §4. The book by Hinze (1959) gives a comprehensive account of the structure of turbulence.

The flow turbulence eddies can be considered as vector velocities in the X and Y directions perpendicular to the flow axis and the Z direction along the flow axis (figure 2). The three component velocities X, Y and Z can be measured in the laboratory by laser velocity meters or by hot wire probes. They are all stationary ergodic variables as shown in figure

3(a). The mean values of the X and Y components are zero, because there can be no flow of the fluid through the pipe wall. The mean value of the Z component is of course equal to the transport velocity of the fluid. An 'observer' moving at the same velocity as the flowing fluid would note that the eddies at position A gradually changed their shape as they moved down the pipe, however; provided that the downstream sensing point B is reasonably close to A (say less than five pipe

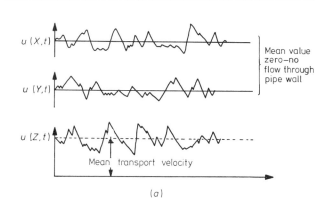

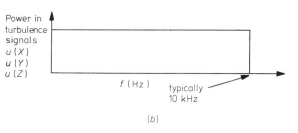

Figure 3 (a), Turbulence vectors are stationary ergodic signals; (b), power spectrum of turbulence signals.

diameters), the eddies are sufficiently similar to be recognised by the correlator in order to enable the transit time to be measured. Clouds due to atmospheric turbulence behave in a similar manner, they keep a reasonably recognisable shape over a large arc of the sky so we can estimate wind speed from them.

The turbulent eddies can be represented by their frequency spectrum, figure 3(b). The spectrum is a band-limited white noise process, with its upper frequency dependent on viscous damping of the turbulent eddies. Typically the upper frequency limit may be as high as 10 kHz. Ideally the flow sensors in a cross correlation flowmeter should respond to the whole range of these turbulent frequencies, this would give the maximum statistical data into the correlator and enable accurate flow measurement with a very short response time. Unfortunately

this wide bandwidth is only possible when using point sensing devices such as laser probes, hot wires etc., which are much too delicate for use in industrial environments. The sensors for industrial use, such as ultrasonic beams, capacitance and conductivity sensing plates etc., interrogate (spatially average) a substantially sized volume of the flow, and hence they will not respond to the higher frequency turbulent eddies. A prime need in system design is therefore to consider the averaging effects of the sensors used, because these fundamentally limit the information bandwidth to the cross correlator and hence limit the accuracy and response time of the system.

2.2 *Spatial averaging effect of the cross correlation sensors*
Referring to figure 2 the spatial averaging effect can be defined as a transfer function relating the turbulence $u(X, Y, Z, t)$ at a point in the pipe to the signals $m(t)$, $n(t)$ into the sensors. The calculation of this relation is complex for sensors having a non-uniform field, such as capacitance and electrical conductivity sensors. Even with ultrasonic sensors which can focus beams across the pipe, the calculation is complicated by the effects of beam reflection and standing waves.

The following analysis will be based on the use of optical sensors which detect the absorption of a beam of parallel light by solid particles in the flowing fluid. Optical beams are well defined so that the analysis is simplified. Even though optical transducers are not often used in industry (because of problems with soiling of optical surfaces) the following model will serve to illustrate the major factors involved in designing any type of sensing system.

Referring to figure 4, turbulently developed clouds of particles pass at velocity u through an optical beam of diameter γ. The incident light I_1 lumens is attenuated by the particles to allow the emergent light I_2 to pass to the receiving

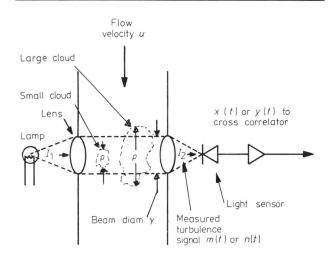

Figure 4 Optical sensing system.

93

optics. The emergent light I_2 corresponds to the input signals $m(t)$ and $n(t)$ to the sensors shown in figure 2.

Now consider a turbulent cloud of size ρm which is smaller than the beam diameter γ. In this case the emergent light is –

$$I_2 = I_1[1 - (\rho/\gamma)^2 r] \qquad (2)$$

where r is the attenuation coefficient of the cloud.

Any cloud which is larger than the beam diameter $(\rho > \gamma)$ will for a period of time obscure the beam so that–

$$I_2 = I_1[1 - r]. \qquad (3)$$

Examination of equations (2) and (3) suggests that a cloud of size $\rho > \gamma$ is fully detected, but small clouds of size $\rho < \gamma$

have relatively little effect on the emergent light which is received by the photodetector. Now a cloud of size γ m travelling at a velocity of u m s^{-1} will take γ/u s to pass across the beam. Hence the highest unattenuated turbulence signal which can be detected corresponds to a signal having a period γ/u s as shown in figure 5(a). This signal has a fundamental frequency of $1/(2\gamma/u)$ Hz.

Summarising the above argument, the cut-off frequency (f_1) at the averaging volume is

$$f_1 = \frac{1}{2\gamma/u} = \frac{1}{2}\frac{u}{\gamma} = \frac{1}{2}\frac{\text{velocity }(u)}{\text{length }(\gamma)}$$

e.g. for a liquid flowing at 4 m s^{-1}, sensor beam diameter 20 mm, cut-off frequencies $f_1 = 4/(2 \times 2 \times 10^{-2}) = 100$ Hz, the corresponding frequency response of the averaging volume is shown in figure 5(b).

2.3 Frequency response of sensor electronics

The sensor electronics can be easily designed to have a wider bandwidth than the spatial averaging bandwidth derived above. Taking the particular system discussed above, which gave a spatial averaging bandwidth of 100 Hz, it would be reasonable to make the bandwidth of the sensor electronics say 500 Hz. This would allow the sensors to detect all the turbulent information and allow for higher frequency signals, which would be generated if the flow velocity of the fluid increased. A frequency response above 500 Hz may allow

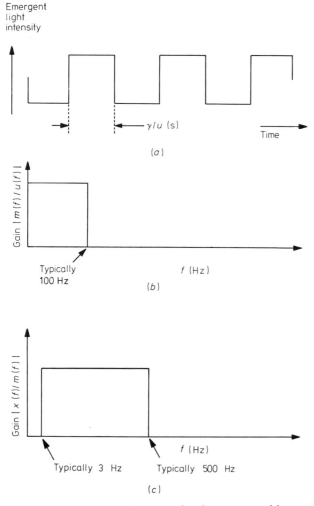

Figure 5 (a), Highest frequency signal unattenuated by spatial averaging; (b), frequency response of averaging volume; (c), frequency response of sensor electronics.

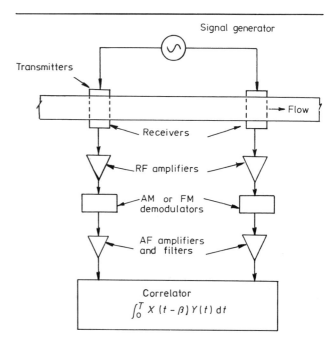

Figure 6 Ultrasonic correlation flowmeter.

unwanted electronic noise to enter the cross correlator, which would reduce the statistical accuracy.

A typical response for the sensor electronics is shown in figure 5(c). The low frequency cut-off point is usually arranged at about 3 Hz so that low frequency signals which cannot be considered statistically stationary within the correlation averaging time (typically 2 s) are not passed to the cross correlator.

3 Sensing the tagging signal

The sensors detect the tagging signal at points A and B, which are located along the flow axis as shown in figure 1.

The candidate sensing principles fall into three broad categories: modulation of radiation by the flowing fluid, emission of radiation by the flowing fluid, electrical and thermal measurement of properties of the flowing fluid. Often several alternative techniques could be used for a practical measuring problem, the following descriptions present various sensing devices to help the reader make an appropriate choice for a particular situation. The choice should be principally based upon reliability and cost of the sensing devices. The gain stability of the sensors is not important, because the cross correlator simply measures the time delay of signals between the sensors, and this time delay is not dependent on the gain.

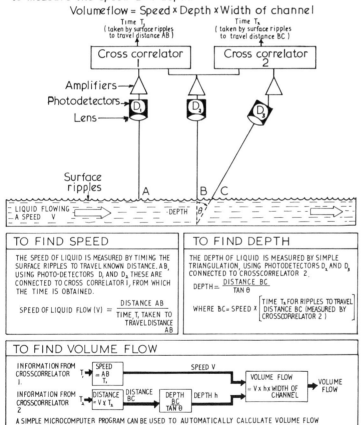

Figure 7 Optical flowmeter for open channels.

3.1 *Modulation of radiation by the flowing fluid*

Radiation techniques are attractive because they can interrogate the flow across a large part of the pipe section, radiation sensors can frequently be mounted externally to the pipe.

(i) *Modulation of acoustic radiation*

When a beam of ultrasound (figure 6) is passed across a flowing fluid, the received ultrasound is modulated by the turbulent eddies in the fluid (doppler effect), or by absorption and scattering effects if particles are entrained in the fluid (Coulthard 1973, Beck and Ong 1975b). For liquids the transmitting and receiving sensors can be mounted on the outside wall of the pipe, frequencies in the region of 1 MHz can pass readily through the pipe wall and this enables a compact and easily maintainable instrument to be constructed, that is likely to find widespread industrial application. With gases the problem is not so simple, the acoustic mismatch between the pipe wall and the low density gaseous medium leads to the requirement of a transducer being in contact with the gas; although ultrasonic cross correlation flowmeters for gases have been successfully demonstrated under laboratory conditions it may be some time before they find their way to industrial development.

(ii) *Modulation of light*

Optical devices find only a very limited application for fluid flow in pipes because of soiling of the optical surfaces. However, optical cross correlation devices have proved successful for measuring flow in open channels using the system illustrated in figure 7, in this case the optical device is mounted well above the flowing fluid and the optical windows can be cleaned by wind-screen wipers if required (Beck and Kaghazchi 1977). Another successful application of optical cross correlation is in measuring the flow velocity of gases from chimney stacks. When steam or particles are present in these gases, turbulent clouds are generated and the velocity of the gas can be measured from a remote point by cross correlating the output signals from two photodetectors which view through telescopes two positions along the axis of the plume (Llewellyn 1976).

(iii) *Modulation of gamma rays etc.*

When solid particles are present in a fluid flow, they can be used to modulate a beam of radiation which is directed perpendicular to the flow. If two such sensors are spaced axially along the flow the received radiation signal can be cross correlated to give the flow velocity. Unfortunately rather strong sources of radiation are required, because it is essential for the count rate registered by the ionising radiation detector to be at least an order above the period of the highest frequency signal to be correlated (Stamford 1980). Hence these techniques may only be worth considering in situations where other devices could not be used, as might arise in measuring the flow in pipes located inside power station boilers etc.

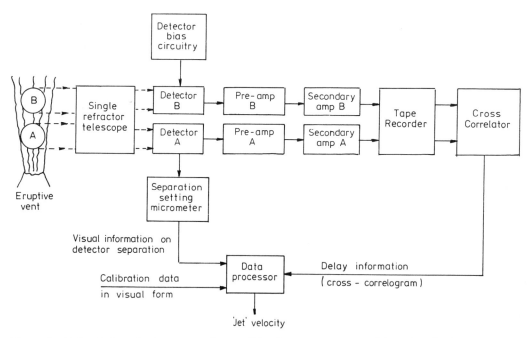

Figure 8 Volcanic jet velocity – antarctic expedition instrument system (from Llewellyn).

3.2 *Emission of radiation by the flowing fluid*

(i) *Thermal radiation*

The most significant applications of radiation techniques have used the thermal radiation emitted by hot gases. The turbulence in hot gas flow inevitably leads to spatial variations of temperature and these provide good tagging signals. They can be detected by infrared sensors using suitably focused telescopes and located at a safe distance from the hot flowing gas.

Referring to figure 8, our first application was concerned with measuring the velocity of hot gas from volcanic jets (Llewellyn 1976). The telescope for these measurements was located over 200 metres from the gas jet and well away from the danger zone caused by the fall out of hot rock and magma which was also emitted from the jet. A similar application (figure 9) was by British Steel who measured the flow of hot gases from oxygen steel-making vessels (Webb *et al* 1976). This latter technique seems to work well and will almost certainly lead to less pollution, better quality steel and better operating efficiency in steel production plants.

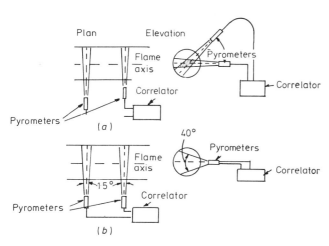

Figure 9 Flow of hot gases from oxygen steel-making vessels (from Webb *et al*). (*a*), Velocity profile sighting; (*b*), mean velocity sighting. The shaded area represents the correlation area.

(ii) *Ionising radiation*

A technique is being investigated (Stamford 1980) for measuring the flow of highly radio-active liquor in nuclear fuel separation plants. Variations in the emitted radiation are caused by either solid particles or gas bubbles in the flowing fluid and it is thought that these variations could be detected by scintillation detectors mounted outside the primary radiation shield, but viewing the pipe through very small holes in the shield. The net effect should be a flowmeter without

parts inside the radiation zone where maintenance would be impossible. The scintillation detectors would be located in bulges to which standard maintenance procedures apply.

3.3 *Instantaneous measurement of electrical and thermal properties of the fluid*

Sensors for these instantaneous measurements have proved simpler to develop than the radiation type devices described above, hence there is already considerable information known about their performance.

(i) *Capacitance sensors*

Capacitance sensors are particularly useful for measuring the flow of gases with entrained solids. The solids form clouds due to turbulence and these clouds modulate the capacitance of a plate which is exposed to the flowing fluid. By cross correlating the capacitance modulation of two plates, suitably spaced along the axis of the pipe, the flow velocity can be measured (Beck *et al* 1968, Kippham and Mesch 1978). The capacitance sensing electrodes (figure 10) are very hard wearing, they do not protrude into the flow and the necessary insulating materials are available to withstand most conditions of temperature, corrosion or abrasion which may be encountered. Capacitance sensors have been used for several years in shot blasting machines, where they have been exposed to extremely abrasive conditions with no deterioration in performance (Beck *et al* 1978).

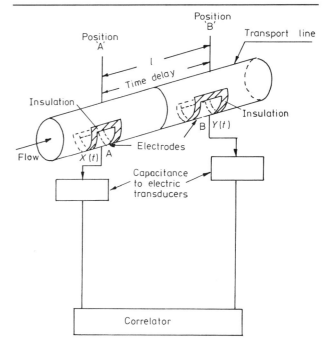

Figure 10 Schematic diagram of pneumatic conveyor, transducers and cross correlator.

(ii) *Electrodynamic sensors*

Whenever solid particles are entrained in a flowing stream of gas a considerable electrostatic charge accumulates on the particles. This electrostatic charge can be detected by sensors using plates somewhat similar to the capacitance sensors described above. The voltage on the plate is amplified by a relatively simple AC amplifier and the output can be cross correlated to give the flow velocity (King 1973). In many ways they are similar to the capacitance sensors, but it should be possible to make them even more sensitive for measuring the flow of gases with very small amounts of entrained solids; such situations are likely to occur in suction lines feeding industrial gas cleaning plants.

(iii) *Electrical conductivity sensors*

The electrical conductivity sensors (Beck *et al* 1974) for liquid/solid mixtures and for liquid/liquid mixtures are analogous to the capacitance sensors for gas/solid mixtures. The discontinuous phase in the liquid causes instantaneous variations in the electrical conductivity of an electrode exposed to the flowing fluid, and these changes can be measured and cross correlated to give the flow velocity. The electrodes can be very robust (figure 11); the electronic circuits are very simple and hence these devices could be a strong contender for use as a low cost flowmeter for slurries. They do not have the advantage of clipping on the outside of the pipe, which ultrasonic sensors do, and they will not work reliably if the slurry contains a large amount of grease (as in some sewage slurries) because the grease will deposit on the conductivity electrodes and stop them working. On the other hand they are very simple and will certainly be much cheaper than ultrasonic systems.

(iv) *Electrolytic and injected bubble cross correlation sensors*

Attempts have been made to modify the electrical conductivity

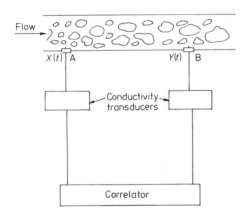

Figure 11 Conductivity cross correlation flowmeter.

sensors so that these will work with single phase liquids. The first attempts used a DC measurement of electrical conductivity. Referring to figure 12, a polarising layer of gas bubbles was formed on the conductivity electrode and these bubbles were swept away at a variable rate according to the velocity of the turbulent eddies as they passed the electrode. Hence a variable current passed into the electrode to replenish the polarising bubble layer and this current was able to be cross correlated to give the velocity. Unfortunately all the practical realisations of this system suffered from chemical deposition on the electrode which eventually stopped it working. More promising, however, is a recent development (Kirkby 1979) in which the electrode is replaced by a stainless steel needle with a very small flow of fine bubbles emerging from its end (figure 13). These bubbles behave in the same way as the polarisation layer mentioned earlier, but in this case the electrode itself can be energised by an AC conductivity meter and it appears that the problem of chemical deposition on the electrode is overcome.

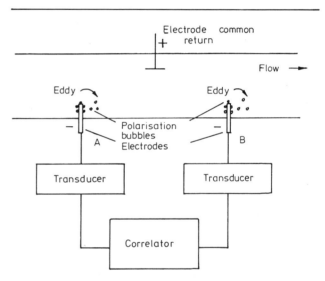

Figure 12 Electrolytic correlation flowmeter (from Kirkby).

(v) *Thermal sensors to detect turbulent eddies*

Turbulent eddies in fluids cause corresponding random variations in the rate of heat exchange with a heated body. A heated thermistor provides a convenient way of detecting these eddies. The thermistor can be run at constant voltage, or constant current, and the corresponding variations in current or voltage due to the temperature changes can be cross correlated. Alternatively the thermistor can be operated as a constant temperature anemometer by including it in a suitable feedback circuit, so that its resistance is maintained constant by varying the current through it; the voltage signal across the

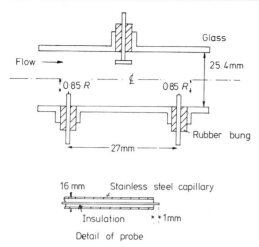

Figure 13 Probe arrangement for electrolytic flowmeter (from Kirkby).

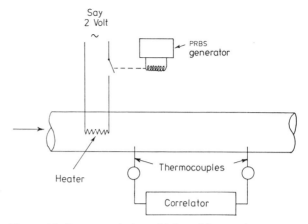

Figure 14 Cross correlation of injected heat pulses.

thermistor can then be used as an input to the cross correlator. It is claimed (Goh and Bentley 1977) that the constant temperature anemometer mode reduces to some extent the thermal time constant (and hence limitation in frequency response) which is inherent in thermistors used in the constant current or voltage modes. The thermistor is probably one of the very few alternatives to the ultrasonic system for a single phase fluid. It is a cheap and simple device to use, but it does intrude into the flow and it does suffer from a limited frequency response unless a very small thermal mass device is used, unfortunately the small thermal mass devices would be too fragile for industrial flows.

(vi) Thermal sensors to cross correlate injected heat pulses
Referring to figure 14, when a fluid is flowing under laminar conditions it may be necessary to inject a tagging signal into the flow (Beck and Abeysekera 1972). This can be done by switching on and off in a pseudo-random manner a small heater in the pipe. The resultant temperature pattern in the flow is detected by thermocouples or thermistors etc. at two points along the flow axis; the detectors outputs are cross correlated to give the flow velocity. When the fluid happens to pass over a heat exchanger, sufficient temperature variations for cross correlation purposes may be generated by the natural randomness of heat exchange. This has been exploited to measure the water circulation of internal combustion engines and to measure the coolant flow in the tubes of a nuclear reactor (Bentley and Dawson 1966). A paper by Boonstoppel *et al* (1968) evaluated the potential accuracy of the technique.

4 On-line cross correlation computation

The calculation of the cross correlation function involves the delaying of one signal with respect to the other, the multiplying of the two signals together, and the averaging of the product over a suitable period of time. The continuous (direct correlation) equation for the above operation is:

$$R_{xy}(\tau) = \frac{1}{T} \int_0^T x(t-\tau) y(t) \, dt. \qquad (4)$$

A digital cross correlation computes the sampled data form of the above equation:

$$R_{xy}(j) = \frac{1}{N} \sum_0^N x(k-j) y(k). \qquad (5)$$

Initially an expensive general purpose digital computer was used to perform this correlation. An analogue correlator using a magnetic tape delay was next used (Butterfield *et al* 1961) then a general purpose digital correlator (Anderson and Perry 1969). All these systems have been too expensive for general on-line industrial use.

The data sampling rate for flow cross correlation can be as high as 1 kHz. At this speed many digital computers and currently available microprocessors are unable to solve equation (5) in real time. A reduction in the amount of computation required can be obtained by cross correlating only the polarity of the input data and solving the equation:

$$R_{xy}(j) = \frac{1}{N} \sum_0^N \text{sgn } x(k-j) \text{ sgn } y(k) \qquad (6)$$

where sgn denotes the sign of the data with respect to its mean value. In a polarity cross correlator the data is read as a series of binary values as shown in figure 15. By reading binary values the amplitude information in the data is lost and the statistical accuracy of the cross correlation will suffer. This can be compensated for by increasing the data observation time by a factor of about 2·2 over the time required for the direct solution of equation (5) (Jordan 1979).

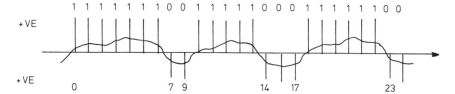

+VE

+VE

1 1 1 1 1 1 1 0 0 1 1 1 1 1 0 0 0 1 1 1 1 1 1 1 0 0

0 7 9 14 17 23

Figure 15 Zero crossing encoding.

In order to avoid this approximate doubling of response time necessary with a polarity cross correlator, 'relay' cross correlators have been designed in which one data stream is polarity quantised and this is correlated with the actual value of the other data stream using the equation

$$R_{xy}(j) = \frac{1}{N} \sum_{0}^{N} \text{sgn } x(k-j)y(k). \tag{7}$$

Relay correlation requires the data observation time to be increased by a factor of 1·4 over the time required for the direct solution of equation (5).

4.1 Correlators designed specifically for flow measurement
(i) Peak seeking optimisers
Early designs of flow cross correlators used a hill climbing arrangement (Mesch *et al* 1971) so that the correlator time delay was automatically adjusted to give the maximum output of the cross correlation integrator as shown in figure 16. This arrangement appears attractive for minimising the hardware required, but unfortunately when the flow changes suddenly, or when there are certain disturbances in the flow, there are small spurious peaks on the cross correlation function and the peak-seeking circuit can lock on to a subsidiary peak, giving a gross error in flow measurement. This could be overcome by using logic circuits which direct the correlator to periodically rescan the whole function and reselect the major peak. These additional circuits increase the cost and complexity to a level where it seems preferable to use special large scale integrated circuits or microprocessors to inspect the whole cross correlation function and obtain the highest peak.

(ii) Special purpose large scale integrated circuits
This system uses the fact that the correlation function grows with increasing integration time (figure 17). If an arbitrary level is chosen, and the correlation function is allowed to grow through this level, then the first part of the function to reach this level is the peak. This principle can be implemented in a circuit having a regular repetitive form that can be readily realised in an integrated circuit (Jordan and Kelly 1976).

(iii) Microprocessor cross correlators
Microprocessors are attractive because they are cheap and becoming even cheaper. They are too slow to solve even the polarity cross correlation in real-time, but a number of systems using external circuits for delay and high speed multiplication have been described (Jordan 1979, Coulthard and Keech 1979). Unfortunately these add complications and expense over the attractive concept of a single chip microprocessor correlator. This problem was considered by Henry who devised a new cross correlation algorithm for a microprocessor (Henry 1979) which increased its speed by well over ten times and eliminated the need for the additional circuits.

Henry's arrangement is based on the principle that the only

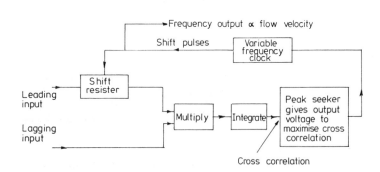

Figure 16 Peak seeking cross correlator.

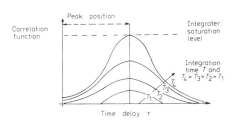

Figure 17 Peak detection in LSI correlator chip.

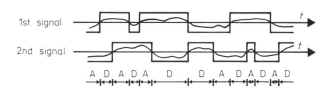

Figure 18 Cross correlating from zero-crossings. A, agreement; D, disagreement.

significant information used by a polarity cross correlator is the actual zero crossing times of the signals as shown in figure 18. This information can be stored in the computer as binary data denoting agreement or disagreement between the signals as shown. This avoids the need for sampling the data at close intervals that were previously thought necessary to give adequate resolution (say 0·2%) in flow measurement as shown in figure 15. Henry noted that the zero crossings are randomly distributed in time, hence the observation by the correlator of their time of occurrence gives data with a very fine time resolution. This seems to be an optimal way of handling polarity data and accounts for the large savings in computer time.

4.2 *Future developments in flow cross correlators*
In the immediate future it seems almost certain that most requirements will be met by microprocessor cross correlators using polarity correlation and based on Henry's method for maximising computer efficiency.

Polarity correlation (equation (6)) does have the penalty of slower response time compared with direct correlation (equation (5)). Intermediate systems using two or three level amplitude quantisation could speed up the response without excessive computational effort. However, looking further into the future one can see the possibility of microprocessors being constructed using charge coupled devices, storage and computation logic in a single package (Arthur *et al* 1979). The charge coupled device would offer the advantage of analogue data handling and would give the possibility of making correlators with the best possible response time at a realistic cost for industrial flow measurement.

5 Accuracy of cross correlation flowmeters
The measured flow rate is obtained using the flow model (equation (1)) derived in §2.1. The measured length (l) and measured cross sectional area (A) are subject to normal geometrical constraints. However, uncertainties in the measurement of the time delay τ^* and the calibration factor k affect the flow measurement accuracy in a more complex way.

5.1 *Calibration factor k*
It is convenient to consider the calibration factor k as a product of three factors, k_1, k_2 and k_3

$$k = k_1 k_2 k_3$$
$$k_1 = \text{flow profile factor}$$
$$k_2 = \text{velocity/time trade-off factor}$$
$$k_3 = \text{sensor phase error factor.}$$

Now examining these factors individually

(i) *Flow profile factor k_1*
The flow profile across a pipe is not uniform. An ideal sensor would take an average of the turbulence information across the whole cross section of the pipe, hence one would expect the value of k_1 to be equal to unity. Unfortunately the sensors currently available do not average the whole cross section, for example an ultrasonic beam averages over a diameter of the pipe and for turbulent flow geometrical analysis shows that a flow profile correction factor of $k_1 = 0\cdot94$ would be required to allow for the effects (Beck and Ong 1975a).

(ii) *Velocity/time trade-off factor k_2*
This factor allows for the basic fact that eddies in the flow which travel most quickly (usually those near the centre of the pipe) have less time to decay between the sensing zones A and B (figure 2) than have the slower moving eddies nearer to the wall of the pipe. This causes the measurement to be biased towards the faster moving eddies. We have not been able to determine the factor k_2 from theoretical considerations, but we think it may be sufficiently close to unity to be reasonably confident that it can be determined when calibrating the flowmeter, and that its value will not change significantly in service.

(iii) *Phase error in sensors*
Sensors which instantaneously interrogate the flow (e.g. optical, capacitance, conductivity etc.) can be designed to cause virtually no problems with phase errors. All that is necessary is for both the sensors to be reasonably well matched.

A potentially serious problem can be caused by mismatch of the acoustic beams when using ultrasonic sensors. The problem here is that reflections at the fluid to pipe wall interface cause a high standing wave ratio (say a ratio of 5/1) across the acoustic cavity in the pipe; the turbulent eddies are detected by their modulating effect on the standing wave pattern. If the standing wave patterns are mismatched, by say

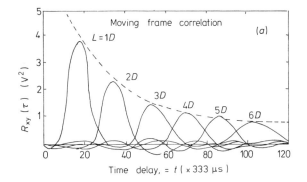

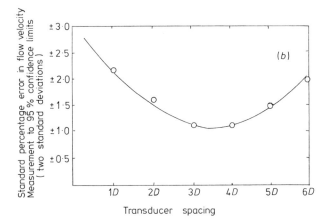

Figure 19 (a) Measured cross correlation function against transducer spacing; (b) normalised standard error in flow velocity measurement against transducer spacing. Experimental data: water flow in one inch (25·4 mm) diameter copper pipe. Ultrasonic probe diameter = 10 mm; carrier frequency = 1 MHz; AM signal processor filter bandwidth = 1–5 kHz; integration time = 20 s; mean velocity V_m = 4·2 m s^{-1}; L = transducer spacing; D = pipe diameter (25·4 mm).

a half wavelength, the cross correlation peak will be inverted. If mismatched by a quarter wavelength the peak will be shifted by a 90° phase shift. The resultant errors would be intolerable, so it is essential to match the standing wave patterns to within a few degrees. Now for measuring liquid flows ultrasound frequencies as high as 1 MHz have to be used for the ultrasound to penetrate the pipe wall. At this frequency the acoustic wave length is about 1·5 mm. Even with the best physical and electronic matching of the sensors, it would not

be possible to maintain the standing wave pattern sufficiently constant for long periods of time in a practical process situation. Fortunately, recent work indicates that this problem can be overcome by using phase control schemes and adaptive frequency trackers to control precisely the phase and acoustic wave pattern across the cavity (Leach 1978).

5.2 *Errors in determining cross correlation time delay* τ^*

The cross correlator itself is a digital device, with its timing controlled by a crystal clock. Therefore there is no significant fundamental error in the cross correlation measurement of time delay. There could be errors due to quantisation of the time delay measurements by the digital device, however, these errors can be overcome by suitable design of the digital system (Jordan and Kelly 1976).

A fundamental source of statistical error is the limited bandwidth of the signals input to the correlator and the limited time available to make a measurement. In considering these limitations, the need for flow measurement can be sub-divided into two separate areas:

(i) Total flow measurement for accounting and custody transfer purposes etc. In such cases the averaging time of the flow measurement is very long and there would be negligible statistical error provided that the flow is steady during the averaging time of the cross correlator. If the flow varies, the averaging time may have to be quite short and therefore the accuracy considerations discussed in (ii) below will apply.

(ii) Process flow control where the integration time of the cross correlator should be between say 1 and 5 s in order to enable the control loop to control against load changes. In these cases the statistical error could be significant and this will be discussed below to show which of the system parameters can be adjusted to minimise the error.

5.2.1 *Statistical errors in cross correlation flowmeters.*
The standard error $\epsilon(\tau)$ of the transit time measurement is a function of signal bandwidth (B), correlator integration time (T) and the normalised cross correlation (ρ^*). Several investigators have developed relationships of the form (Beck and Ong 1975a):

$$\epsilon(\tau) = \frac{k_4}{B^{1\cdot5}T^{0\cdot5}} [1 + (1/\rho^*)^2]^{0\cdot5} \qquad (8)$$

where k_4 is a constant.

The flow rate Q is related to the transit time measurement as shown in equation (1), hence the standard error in flow measurement $\epsilon(Q)$ is given by:

$$\epsilon(Q) \propto \epsilon(\tau)/l. \qquad (9)$$

Now turbulence decays as the length l increases, hence the standard error $\epsilon(\tau)$ (equation 8) increases in a nonlinear way as the length l increases. Equation (9) shows that increasing the length l can reduce the error in flow measurement. The overall result is that excessively short length (very short transit time) and excessively long length (poor correlatability of signals) can result in poor flowmeter performance, but there is a range of values of between say one pipe diameter and six pipe diameters which gives an acceptable performance. The way in which the cross correlation function decays as l is increased is shown in figure 19(a) and the corresponding way the standard error varies is shown in figure 19(b).

6 Conclusions

We have seen how cross correlation flowmeters illustrate the advantages of the use of correlation for measurement purposes. The principle features being firstly that calibration of the sensors is not required because the time delay is measured with reference to a crystal controlled time standard in the cross correlator, secondly cross correlation rejects the effect of spurious interference on the signals. This is clear when observing the records from the two sensors of a cross correlation flowmeter, the turbulence pattern breaks up between the sensors so that the eye cannot determine the time delay relationship, fortunately the cross correlator can recognise and precisely determine this time delay.

We can expect cross correlation flowmeters to be developed to solve a wide range of industrial and environmental flow measurement problems. Until recently the commercial development of these flowmeters was inhibited by the high cost and complexity of cross correlators. We have seen how recent developments in large scale integrated circuits and microprocessors should enable simple, fast and reliable cross correlators to be designed at a cost that should be acceptable for industrial use. A few of the possible applications of cross correlation flowmeters are being developed by industry, but undoubtedly there is still a considerable amount of basic research work to be done concerned with sensor design and obtaining a better understanding of the way in which the sensors interrogate the flow.

In future, new cross correlation measurement systems will be developed and some of these will make even greater demands on data processing speed, fortunately we can expect recent developments in data processing such as the use of charged coupled devices (Arthur et al 1979) and optical signal processing (Murison et al 1979) to provide the required increase in data processing speed.

Acknowledgments

I am grateful to all the Bradford University students, staff and technicians and also my industrial colleagues whose enthusiasm and painstaking effort has enabled the work to be done. I also thank the Science Research Council for providing so much of the money needed for the work and the National Research Development Corporation for guiding the applications of the techniques.

References

Anderson G and Perry M A 1969 A calibrated real time correlator/averager/probability analyser
Hewlett Packard Journal, vol 21, no 3, pp 9–15

Arthur J W, Dix J F, Harland E, Widdowson J W, Denyer P B, Mavor J and Milne A D 1979 Large time-bandwidth product CCD correlators for sonar
IEE Conference Publication 1979/180 pp 62–68

Atomic International 1965 Development of noise-analysis techniques for measuring reactor coolant velocities
NAA-SR-11193

Beck M S, Drane J, Plaskowski A and Wainwright N 1968 Particle velocity and mass flow measurement in pneumatic conveyors
Powder Technology **2** 269–77

Beck M S and Abeysekera S A 1972 Liquid flow measurement by cross correlation of temperature fluctuations
Measurement and Control **5** T143–7

Beck M S, Lee K T and McKeown K J 1974 An on-line instrument for measuring small quantities of dispersed non-conducting liquid in a conducting liquid
Measurement and Control **7** T65–9

Beck M S and Ong K H 1975a Volume flow measurement of liquids and slurries by cross correlation of natural turbulence signals
Proc. of Conference on Fluid Flow Measurement in the Mid 1970's National Engineering Laboratory, April 1975

Beck M S and Ong K H 1975b Slurry flow velocity, concentration and particle size measurement using flow noise and correlation techniques
Measurement and Control **8** 453–63

Beck M S and Kaghazchi B 1977 Remote level and velocity measurement of rivers and open channels by analysis of random signals from surface waves
IEE Conference Publication No 159, April 1977

Beck M S, Green R G, Kwan H K and John R 1978 A low cost instrument for solids flow measurement in industry
J. Phys. E: Sci. Instrum. **11** 1005–10

Bendat J S and Piersol A G 1971 *Random Data: Analysis and Measurement Procedures* (New York: Wiley)

Bentley P G and Dawson D G 1966 Fluid flow measurement by transit time analysis of temperature fluctuations
Trans. Soc. Instrum. Technol. **18** 183–93

Boonstoppel F, Veltman V and Kergrouwen F 1968 The measurement of flow by cross correlation techniques
IEE Conference, Publication no 43, June 1968

Butterfield M H, Bryant G F and Dowsing J 1961 A new method of strip speed measurement using random waveform correlation
Trans. Soc. Instrum. Tech. **13** 111–23

Coulthard J and Keech R P 1979 A six channel microprocessor controlled correlator
IEE Conference Digest no 1979/32 pp 4/1–6

Cousins T 1978 The doppler ultrasonic flowmeter
FLOMEKO 78 (Amsterdam: North-Holland) pp 513–8

Davies D E N 1979 Signal processing for radar applications
IEE Conference, Publication no 1979/180 pp 145–50

De Jong J 1978 Comparisons of some 500 mm diameter electromagnetic flowmeters
FLOMEKO 78 (Amsterdam: North-Holland) pp 209–13

Durrani T S and Greated C A 1977 *Laser Systems in Flow Measurement* (New York: Plenum Press)

Dyer D and Stewart R M 1977 Detection and rolling element bearing damage by statistical vibration analysis
Symposium on On-line Surveillance and Monitoring of Process Plant (London: The Society of Chemical Industry) pp 3.1–3.13

Eykhoff P 1974 *System Identification* (London: Wiley)

Fischer M J and Davies P O A L 1964 Correlation measurements in a non-frozen pattern of turbulence
Journal of Fluid Mechanics **18** 97–116

Goh C K and Bentley O P 1977 Investigation into cross correlation techniques to measure low velocity flue gases
FLOW-CON 77 (London: The Institute of Measurement and Control) pp 409–32

Hayes A M 1976 *PhD Thesis* School of Electrical Engineering, Bradford University

Henry R M 1979 An improved algorithm allowing fast on-line polarity correlation by microprocessor or minicomputer
IEE Conference Digest no 1979/32 pp 3/1–4

Hinze J D 1959 *Turbulence* (New York: McGraw-Hill)

Idowaga T and Ono T 1976 A new method of vehicle-speed measurement using cross correlation
Proc. IMEKO VII, London, May 1976 (London: Institute of Measurement and Control) pp BLV 218/1–9

Jordan J R 1979 Correlation circuits for measurement systems
IEE Conference Digest no 1979/32 pp 1/1–4

Jordan J R and Kelly R G 1976 Integrated circuit correlator for flow measurement
Measurement and Control **9** 267–70

Kashiwagi H 1968 Paper strip speed measurement using correlation techniques
Trans. Soc. Instrument and Contr. Engrs., Japan **4** 304–12

King P W 1973 Mass flow measurement of conveyed solids by monitoring of intrinsic electrostatic noise levels
Pneumotransport 2, 5–7 Sept. 1973 (Cranfield: British Hydrodynamics Research Association) pp D2–9

Kipphan H and Mesch F 1978 Flow measurement systems using transit time correlation
FLOMEKO 78 (Amsterdam: North-Holland) pp 409–16

Kirkby H J A 1979 *MPhil Thesis* School of Control Engineering, Bradford University

Krause F R, Betz H T and Lysobey P H 1973 Pollution detection by digital correlation of multispectral, stereo-image pairs
Atmospheric Environment **7** 455–76

Lange F H 1967 *Correlation Techniques* (London: Iliffe Books)

Leach K 1978 *PhD Thesis* School of Control Engineering, Bradford University

Llewellyn G J 1976 Development of an instrument for measuring volcanic jet-velocity
Proc. IMEKO, V. 11, London, May 1976 (London: Institute of Measurement and Control) pp 247/1–6

Mesch F, Daucher H H and Fritsche R 1971 *Correlation Method of Velocity Measurement. Part I, Principles, Application to Two Phase Flows* (Germany: Meestechnik) Vol 79, no 7, pp 152–7 (CEGB Information Services, CE Trans. no 5890)

Murison A, Grant P M and De La Rue R M 1979 Application of optical techniques to signal processing
IEE Conference, Publication no 1979/180 pp 34–45

Open University, *Transducers 2: Flow, Instrumentation Unit 10* (Milton Keynes: The Open University Press) T291/10

Shinners 1973 *Modern Control System Theory and Application* (New York: Addison–Wesley)

Stamford B 1980 *PhD Thesis* School of Control Engineering, Bradford University

Watts F W 1980 Ultrasonic flowmeters: their reliability and accuracy in industrial plants

I. Chem. E. Symposium Series no 60 pp 147–60

Webb P J, Ashton R M and Kenahan T 1976 Remote monitoring of the Properties of Exhaust Fume from Oxygen Steelmaking Processes
Proc. IMEKO, vol 11, London, May 1976 (London:

Institute of Measurement and Control) pp AML/148/1–10

Zimmer C, Ryser H and Meyr H 1976 An instrument for non contact speed measurement by image correlation
Proc. IMEKO, vol 4, London, May 1976 (London: Institute of Measurement and Control) pp BLV/217–1 to 10

Chapter 9

Nucleonic instrumentation applied to the measurement of physical parameters by means of ionising radiation

R B J Palmer

Department of Physics, The City University, Northampton Square, London EC1V 0HB, UK

Abstract. The general principles of nucleonic instrumentation are discussed with reference to the requirements for optimal design. Techniques currently in use for the measurement of thickness, density and other related physical parameters by means of beta, gamma and heavily ionising radiation are critically reviewed.

1. Introduction

During the past twenty years the measurement of physical parameters by means of nucleonic instrumentation has been extended into many fields, and a number of reviews have surveyed the subject. e.g. Gardner and Ely (1967), Cameron and Clayton (1971) and Taylor (1971). New techniques have been devised and the advent of new and improved radiation sources and detectors has widely increased the scope and accuracy of measurements. More recently the rapid expansion of signal processing facilities has made it possible to present the information in various forms which offer considerable advantages for industrial or commercial use. In such applications any advantages in convenience or reliability must, of course, be considered in the context of cost, working conditions and safety, and the practical applications of nucleonic instrumentation therefore range from very simple and rugged systems, which may give all the information required in particular circumstances quite cheaply, to highly sophisticated instruments which are capable of the analysis of complex physical and chemical parameters. In the past, uncertainty and fears concerning the hazards involved in the use of radioactive materials have tended to limit the application of nucleonic techniques. However, adequate legislation and test procedures should restore confidence in these methods which, if rightly used, present no greater hazard than other forms of modern equipment.

The use of readily available sources of alpha, beta and gamma radiations has been exploited for the measurement of physical parameters such as thickness, density, particle size, flow characteristics, etc, in a variety of different ways which have been summarised in figure 1. Recently there has been considerable emphasis on optimisation of design with respect to the choice of sources, detectors, geometry, and general techniques which are most suitable for various purposes, and this should lead to a general improvement in the standard of measurement and in some cases to the possibility of using weaker sources.

In certain applications the use of electrically generated particles can be an advantage. This usually involves greater expense in capital outlay and maintenance but in some cases this can be justified, e.g. where protons and ion beams in general are required machine generation is the only source available.

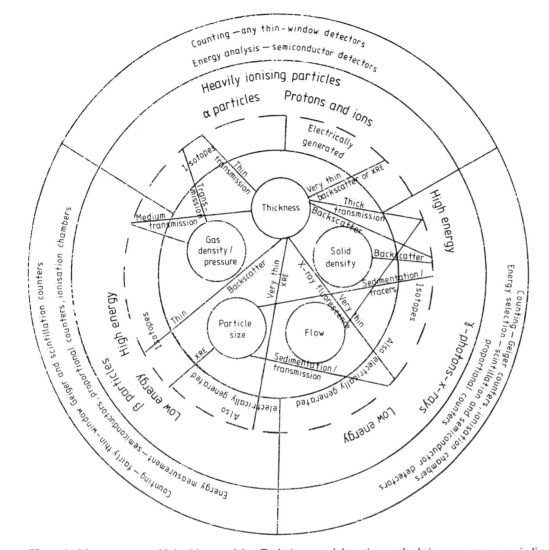

Figure 1. Measurements with ionising particles. Techniques and detection methods in common use are indicated.

2. Basic principles and procedures

The major distinctions that can be made between ionising and non-ionising radiation arise from the fact that the latter, e.g. ultrasonic or long-wavelength electromagnetic radiation, does not produce free electrons in the absorbing medium. Most forms of radiation induce atomic or molecular vibration or electronic excitation, but the free electrons produced in the vicinity of a beam of ionising radiation give rise to distinctive physical and chemical effects and render the radiation very sensitive to detection by many different means. A minimum energy of between 4 and 25 eV (approximately 2×10^{-18} J) is required to ionise an atom, and in practice the energy of the radiation used is considerably greater than this, so that a single particle or photon produces many ion pairs in an absorbing material. The precise mechanism of the interaction with the material depends upon the nature and energy of the radiation and often there are competing forms of interaction. The initial ionisation sometimes produces a relatively high-energy electron, and this primary emission can then give rise to secondary electron emission. Where this occurs the deposition of energy is not localised in the direct path of the incident radiation but may be spread over a relatively large volume. The effect of scattering is marked in the case of beta radiation and is even more pronounced in the attenuation of high-energy gamma radiation, Compton

107

scattering being the dominant form of interaction at certain energies. This releases an energetic electron and a gamma photon of degraded energy.

Owing to the diversity of possible interactions the attenuating effect of different materials varies considerably with the type and energy of the radiation and with the atomic number of the material. Ionising radiation can therefore be used for the measurement of physical parameters in a very wide variety of circumstances, but the type of radiation chosen and the techniques employed must be carefully matched to the conditions.

2.1. General procedures

Measurement of the physical parameters of a material basically requires a source of radiation, a detector and some suitable means of assessing the information received. The relative positioning of the source, material and detector require careful consideration. In some circumstances collimation of the radiation is desirable so that it is incident upon the material over a well defined area. Typical arrangements which rely upon the measurement of transmitted, forward scattered and back-scattered radiation are shown in figures 2(a) and (b). The intensity of scattered radiation is low compared with that of the incident radiation, so when this is being measured it is desirable to collect it through a large solid angle and it is essential to shield the detector well from any background radiation in order to maximise the signal-to-noise ratio.

If good collimation is not used fixed geometry is very important since a small change in area or position of the test material can seriously affect the detector response (figure 3). Shielding can be more difficult under these conditions since scatter is more widespread. The main advantage of the use of wide-angle emission from source assemblies arises from the high proportion of radiation from the source which can be used; collimation severely reduces the effective source strength.

2.2. Sources

Basic information about a number of useful isotopes with fairly long half-lives is listed in table 1. More detailed specifications are to be found in the *Radiation Sources Catalogue* (1981). For most industrial purposes it is desirable to use a long-lived isotope to avoid frequent recalibration and expense. There are exceptions to this, e.g. in flow measurements employing radioactive tracers it is desirable to use fairly short-lived isotopes in order to avoid long-term contamination problems.

2.3. Detectors

Characteristics of the most commonly used detectors are summarised in table 2. All of these rely upon ionising events occurring within a sensitive volume inside the detector. With some detectors, e.g. the Geiger counter, the resulting signal is independent of the nature of the event. With others, e.g. the

proportional counter or semiconductor detector, the signal size can be used as an accurate measure of the energy imparted in the event. The gaseous detectors usually involve relatively large sensitive volumes, whereas the semiconductor detectors have much smaller sensitive volumes, and consequently have different response characteristics. Scintillation counters rely upon the rapid conversion of ionisation energy to light in certain materials. An ionising event is therefore registered as a small flash of light which is detected and amplified by a photo-multiplier. A wide range of scintillation materials are available from which selection can be made to give the most sensitive response to various types of radiation, e.g. sodium iodide for high-energy photons, anthracene or various types of organic scintillator for beta particles. Many types of detector are to some extent light sensitive and need to be shielded from direct light for optimum response, but scintillation detectors must be enclosed in completely light-tight assemblies. They should also be kept cool and be shielded from magnetic fields for maximum signal-to-noise ratio (§ 3.2).

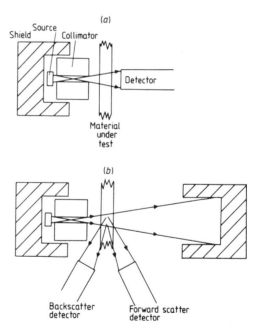

Figure 2. Experimental arrangement for: (*a*), transmission measurements; (*b*) measurements utilising forward or backscattered particles.

The use of solid state detectors has increased recently due to progress in silicon technology. These detectors are small and do not demand high-voltage supplies or complex ancillary equipment for alpha-particle counting or energy measurements where a high degree of precision is not required. With suitable equipment they can, however, give very high resolution in alpha, beta and gamma measurements and it is likely that their use will increase.

2.4. Radiation safety

It must be recognised that there is no 'safe' level of ionising radiation, since any ionising event within the body is potentially hazardous to body cells. However, we are all subjected to a low level of ionising radiation from natural sources, e.g. cosmic radiation, naturally radioactive soil and rock, etc, and also to exposure to radiation used as an aid in medical diagnosis and treatment. This is estimated as being in the region of 3 mGy yr^{-1} (the Gy being the absorbed dose of 1 J kg^{-1}) although this can vary very widely in differing circumstances.

The maximum permissible dose for radiation workers has been devised with the intention that such a dose should not significantly raise the incidence of illness or injury above the level in other industries having a high standard of safety, and the present limit is set at 50 mGy yr^{-1}. This is, however, an oversimplification, since potential effects are linked to the age and sex of the worker, the parts of the body which are subject to exposure, the type of radiation involved and many other factors. Recommendations are made by the International Commission on Radiological Protection, and details are to be found in the *ICRP Publication 26* (1977). Legislation in this country is governed by the Radioactive Substances Acts, 1948 and 1960, and various regulations passed subsequently.

Safety is an important factor in the handling of radioactive isotopes and in the design of equipment associated with them. The main hazard with open sources is ingestion, and they must therefore be handled only by properly trained staff using suitable equipment. Sealed sources must be regularly tested for leakage and adequately protected against theft. Highly active sources are often heavily shuttered except when actually in use, and in some circumstances they are automatically moved from a highly shielded situation into a more exposed position when actually required. In such cases, and also where high levels of x-rays or charged particles are machine generated, access during operation is prevented by a system of interlocking doors in order to prevent the accidental irradiation of personnel. With alpha and many beta sources distance is the best protection, but for gamma sources lead, concrete or heavy-metal shielding is necessary. Amersham International supplies an isotope handling calculator which gives rapid solutions for gamma shielding calculations. It should be remembered that many alpha and beta sources also emit gamma radiation which may present a hazard especially if these are strong sources. The effect of scattering should also not be overlooked in designing shielding for radioactive assemblies. Protection against neutrons is achieved by interposing materials of low atomic number, e.g. water and materials with a high hydrogen and carbon content, and again distance is a good protection.

These precautions may sound formidable, but in situations

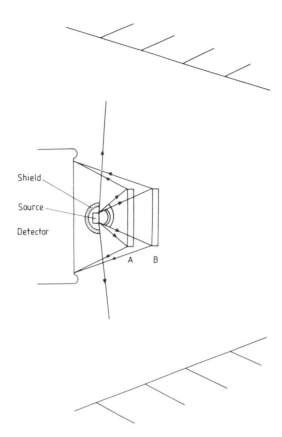

Figure 3. Measurements using a noncollimated source. A small change in area or position of the scattering material, as from A to B, results in an appreciable change in the acceptance angle of particles by the scattering material and by the detector.

where a particular source is involved procedures are well established and, provided these are adhered to, the hazards are negligible. Equipment for monitoring dose levels is readily available at reasonable cost, and there is a film badge monitoring service for personnel who work with radioactive materials. A recent article by Johns (1982) gives an account of radiation hazards and a very useful survey of modern monitoring techniques.

3. Optimum design in nucleonic instrumentation

With nucleonic, as with any form of instrumentation, there are certain general requirements and limitations which may often be conflicting. These involve accuracy, sensitivity, stability, time of response, safety and expense, and the purpose and conditions of use for which the instrumentation is required will determine the relative importance of these factors.

3.1. Accuracy and sensitivity

Accuracy in radiation measurements is ultimately limited by the statistical nature of the emission from radioactive sources. The standard deviation of any radiation count due purely to statistical fluctuations, σ_s, is the square root of the count, $N^{1/2}$, so that a high count rate is obviously desirable for a high percentage accuracy. This can be achieved by having an intense source, a sensitive detector, suitable geometry to give wide angle acceptance into the detector and a long counting time. The intensity of the source that can be used may be limited due to safety considerations, particularly for gamma and neutron sources. It is also limited due to the fact that there will be coincidence losses if the 'dead' time of the detector is not very much shorter than the mean time interval between counts. The length of counting time for any particular measurement is ultimately limited by convenience and stability considerations, and particularly for on-line measurements the counting time for a particular sample may be very short indeed. Good geometrical design and optimum choice of source and detector are therefore essential with source and detector separation as small as the particular application will permit. The overall standard deviation of any count $\sigma_\tau = (\sigma_s^2 + \sigma_i^2)^{1/2}$ where σ_i is the standard deviation arising from all nonstatistical sources of uncertainty, e.g. instrument errors and fluctuations in stability. Where possible σ_i should be appreciably less than σ_s so that $\sigma_\tau \simeq \sigma_s$. For comparative purposes it is convenient to refer to the standard deviation of the count rate, $\sigma_R = (N/t)^{1/2}$.

Nucleonic measurements are concerned with the interpretation of a succession of discrete events and are therefore basically digital in nature. It has often been found convenient, however, to collect the information from these events in an integrated form. For example, in ionisation chamber measurements, the number of electrons collected is registered as a small current flowing through a resistor, and in many applications using Geiger counters the pulses from individual events are collected and smoothed in an RC circuit and

Table 1. Some useful radioisotopes. References: *Radiation Sources Catalogue* (1981), Dyson (1973), Evans (1968).

Source	Predominant energy	Half-life (yr)	Approximate range
α			
^{241}Am	5.5 MeV	433	
^{244}Cm	5.8 MeV	17.8	4.2 to 4.8 cm in air
^{252}Cf	6.1 MeV	2.6	
β†			
^{55}Fe	5 keV	2.7	10^{-3} kg m^{-2} in Al
^{3}H	18 keV	12.3	6×10^{-3} kg m^{-2}
^{63}Ni	66 keV	100	6.5×10^{-2} kg m^{-2}
^{14}C	156 keV	5730	0.3 kg m^{-2}
^{147}Pm	225 keV	2.6	0.5 kg m^{-2}
^{85}Kr	672 keV	10.7	2.4 kg m^{-2}
^{204}Tl	763 keV	3.8	2.8 kg m^{-2}
^{90}Sr	0.55 and 2.3 MeV	28.5	11 kg m^{-2}
γ			$\frac{1}{2}$ thickness in Al
^{3}H/Ti	4.5 keV	12.3	2.5×10^{-2} kg m^{-2}
^{55}Fe	6 keV	2.7	6×10^{-2} kg m^{-2}
^{147}Pm/Zr	15.7 keV	2.6	1 kg m^{-2}
^{238}Pu	17.2 keV	87.7	1.4 kg m^{-2}
^{210}Pb	46 keV	22.3	16 kg m^{-2}
^{241}Am	60 keV	433	25 kg m^{-2}
^{152}Eu(^{151}Sm)	76 keV	87	33 kg m^{-2}
^{147}Pm	121 keV	2.6	46 kg m^{-2}
^{137}Cs	0.66 MeV	30.1	92 kg m^{-2}
^{22}Na	0.51 and 1.27 MeV	2.6	120 kg m^{-2}
^{60}Co	1.17 and 1.33 MeV	5.3	130 kg m^{-2}

† All β sources listed except ^{55}Fe give a broad energy spectrum from zero to the maximum energy stated.

Table 2. Detectors in common use.

	Main applications	Advantages	Limitations
Geiger counters	β and γ counting where discrimination is not required	Cheap. Simple ancillary equipment	Counting only – no energy measurement
Ionisation chambers	Measure total ionisation in fixed gaseous volume	Analogue measurement of high dose-rates	High amplification required. Not suitable for low dose rates
Proportional counters	Energy discrimination	Wide range of discrimination. Can be used for neutrons	Steady high-voltage supply and amplification required
Scintillation detectors	Mainly for γ and β measurements	Sensitive. Can be used for energy measurements	Very light sensitive
Semiconductor detectors	Applicable to α, β and γ measurements	Discriminating. Small. Capable of high resolution	Ancillary equipment can be expensive, especially for high-resolution γ measurements

displayed in analogue form on a rate meter. In this case the rate meter time constant τ should ideally be short compared with the observation time in order that a true response should be obtained; the standard deviation of the rate meter readout R will then be $\sigma_R = (R/2\tau)^{1/2}$. In the case of a change of physical parameter giving rise to a change of true count rate from R_i to R_f, the observation time t_0 required for the individual reading to be within one standard deviation of R_f is given by

$$t_0 \geqslant \tfrac{1}{2}\tau \ln[2\tau(R_f - R_i)^2/R_f]$$

and therefore depends upon the magnitude of the expected change in count rate. If the time constant of the rate meter is too short, however, there is little smoothing of the statistical fluctuations so that assessment of the response is highly subjective. In most instruments the time constant can be adjusted to an estimated optimum value by the operator for a particular count rate. From these considerations it may be seen that if a rapid response is required with a small uncertainty, a high count rate is essential.

A further requirement in the measurement of physical parameters is that a high count rate must be matched by a large change in count rate for a small change in the measured parameter, i.e. for a variable parameter x corresponding to a count rate R, dR/dx should be large for sensitivity in measurement. The standard deviation in x, σ_x, is related to σ_R by $\sigma_x = \sigma_R \, dx/dR$. The fractional change in x which has a 68% probability of being detected is usually referred to as the relative resolving power of the instrument and is given by the relation

$$\Delta x/x = (2^{1/2}\sigma_R/x) \, dx/dR.$$

If a higher confidence level is required this is modified by an appropriate numerical factor.

These considerations are discussed fully by Gardner and Ely (1967) who show that, for an exponential relationship between x and R of type $R = R_0 \exp(-\mu x)$ which is often relevant to beta and gamma transmission measurements in particular, optimum accuracy in measurement is obtained for radiation having an absorption coefficient μ such that $\mu x = 2$ provided that statistical factors are mainly responsible for the uncertainties. Liljestrand *et al* (1976) plot the sensitivity function illustrating the dependence of $\Delta x/x$ upon μx, and it can be seen that this exhibits a shallow minimum for values of μx between about 0.5 and 5. From this the percentage accuracy for a given count rate can be calculated, and it becomes clear that this falls off very rapidly if values of μx are involved which are outside this critical range.

Kurstedt (1976) also discusses optimisation of design and gives various 'rule of thumb' approaches to nucleonic instrumentation which could be useful in the initial stages of feasibility evaluations. Williams (1979) also gives 'tips on nuclear gauging' for level and density measurements mainly applied to civil engineering or large-scale industrial situations.

A more generalised approach is adopted by Notea and Segal (1974a) to the characterisation and optimal design of gauges. Expressions for the resolving power are obtained corresponding to the predicted forms of response of six different gauge systems to the variable under observation, the appropriate types of application being indicated in each case. A systematic procedure is then evolved for the optimisation of design in general, and two particular examples of linear and exponential response functions are discussed.

Segal *et al* (1977) extend these concepts for nondestructive measurements in general and apply the generalised techniques to specific cases of thickness measurement and flaw detection by ultrasonics and by gamma transmission methods. They obtain the optimum resolving powers for measuring the thickness of steel plate by the two techniques as a function of the thickness being measured, thus demonstrating how the most suitable type of instrument for a particular purpose can be selected.

3.2. Signal-to-noise ratio

In any nuclear instrumentation the signal-to-noise ratio must be maximised and in this respect there are many factors to be considered.

Every type of detector has an intrinsic noise level which may depend upon general design factors, thermal effects, sensitivity to fluctuations in voltage supply or various other factors. In choosing the detector for any particular measurement the conditions under which it is to be used must be considered with a view to minimising noise, e.g. the noise level in solid state detectors is very sensitive to temperature, photomultiplier noise is dependent upon thermal and magnetic effects and proportional counters are very sensitive to ripple or slight fluctuations in the high-voltage supply. The characteristics of the particular type of detector should also be studied with a view to maximising the signal, e.g. in matching the ionising radiation to the type and thickness of scintillator and in matching the spectral output of the scintillator to the photocathode of the photomultiplier, and in this respect the general technology of the detector should be carefully studied before finalising the instrument design. Two recent publications on radiation detection by Eichholz and Poston (1979) and by Tait (1980) are helpful in this respect.

There will be an effective contribution to noise due to the presence of background radiation or due to any extraneous radiation accompanying that required for measurement, e.g. unwanted gamma radiation accompanying alpha or beta emission. Suitable shielding, collimation and choice of source and detector can minimise these problems.

The electronic circuitry associated with processing the signals will also make a contribution to noise. Robinson (1974) has written a monograph which discusses noise and fluctuation problems in general. A review of the basic electronics associated with nucleonic measurements together with information about some of the earlier electronic instrumentation is given by Washtell and Hewitt (1965). Nicholson (1974) and Kowalski (1970) have also written specifically on the applications of electronics to nucleonic measurements and Eichholz and Poston (1979) and Tait (1980) have included sections on electronics in their books on radiation detection which describe some of the current techniques in signal processing for nucleonic instrumentation.

3.3. Stability

Allowance must be made for the effect of long-term variations due to instrument changes, e.g. changes in detector characteristics or deterioration of sources, and periodic checking of the calibration is an obvious necessity. In order to counteract the effect of short-term fluctuations in temperature, pressure, voltage supply, etc, comparative techniques are often adopted. Radiation from a single source or from two similar sources is allowed to pass through a standard absorber and through the test specimen, the outputs from each being sampled alternately by a single detector. Any variations in the ambient conditions or in the detector characteristics should then be reflected in both outputs and their effects can be made to cancel. Gardner and Ely (1967) explain how comparative instruments of this kind can be adapted so that the output signals can be used to control on-line processes. Careful consideration must obviously be given to the relative timing of the signal and the control action.

4. Specific techniques

The interaction of radiation with matter is dependent in general upon the number density and nature of the atoms in the path of the radiation. However, the mechanism of the interaction is quite different for radiation of various types and energies. The choice of radiation and the method of application to be used for any particular measurement is therefore very important.

4.1. Heavily ionising radiation

The most commonly used type of heavily ionising radiation in nucleonic instrumentation is alpha radiation since isotopic sources of alpha particles are readily available. These lose energy through a large number of small-loss interactions with electrons in the absorbing medium. Consequently, particles of a precise energy, usually between 5 and 6 MeV, will have a fixed range of less than $50\,\mu\text{m}$ in condensed matter and a correspondingly greater range (approximately in inverse proportion to the density) in gases and vapours. Alpha particle stopping is therefore suitable for the measurement of gas density over a wide pressure range or for thickness measurements of solid or liquid films provided that these are consistent in composition.

If a thin alpha source is used the spread in energy of the alpha particles as measured in a semiconductor detector is very small (1 to 2% spectral energy spread measured at half the peak count). The rate of energy loss in an absorber is in general of the form

$$dE/dx \propto (Z/v^2) \ln(Kv^2)$$

where v is the velocity of the alpha particles, Z the atomic number of the absorber and K a constant dependent upon Z. When the material under test is interposed between the source and the detector the loss in energy can be used as a measure of thickness or density. From the energy loss relation it can be seen that dE/dx increases as the particle energy is reduced so that the response of a detector measuring residual particle energy is most sensitive to density or thickness changes near the end of the particle range. The response of a solid state detector is also temperature dependent, so steady temperature conditions are essential, and unless the source-to-detector distance is very short, air pressure variations can introduce errors.

Thick alpha particle sources, which are available at higher intensities than thin sources, can be used simply in counting mode in conjunction with any thin-window detector. The alpha particles from a thick source cover a wide energy spectrum so that the number of particles reaching the detector will be a function of the absorber density or thickness. By careful source design the alpha spectrum can be controlled so that the response of the detector falls off linearly with density or thickness over a fairly wide range, but calibration is required for use with different materials or gas mixtures. Although not as sensitive to small density variations as the thin source method, a radio gauge designed on this principle requires a simpler apparatus and is versatile in application, and it can be adapted to operate gas pressure control systems.

With alpha particle instruments, a short source-to-detector distance is usually required and consequently the geometrical design is important. Collimation of the particles may be required and this necessarily reduces the effective strength of the source. Geometrical considerations for alpha particle gauging have been discussed by Gardner *et al* (1972a). They have set up theoretical models with noncollimated geometry and tested the predictions experimentally. Their conclusions can be of general use in calibrating and optimising the design of alpha transmission gauges.

A recent alpha particle instrument for the measurement of humidity is described by Matsumoto and Kobayashi (1981) in which a double ionisation chamber is used in order to eliminate the effects of variable temperature and pressure. An accuracy of $\pm 5\%$ is attainable.

Film thicknesses can also be determined by heavy-ion bombardment, estimation of either the backscattered ions or the particle-induced x-ray emission being used for thickness measurement. Ion beam measurements are usually more successful than electron probes which can also be used to induce x-ray emission, because the sensitivity is higher and because beam scattering within the layer is very slight. Ion beam techniques in general are described by Thomas and Cachard (1978) as applied to material characterisation. Thickness measurements have been made in the range 5–200 nm by Kerkow *et al* (1979) with a 280 keV proton beam using a proportional counter as detector. Layers of copper on platinum, carbon and silicon were measured with 10% accuracy, and the method was found to be more reliable than the quartz oscillation method over small areas. It is customary in these measurements to express the thicknesses of films in terms of mass per unit area ($kg\,m^{-2}$ or $g\,cm^{-2}$) since this is the criterion to which the measurement is basically sensitive. This avoids uncertainties due to any density differences in the measured films. Johannson *et al* (1975), using 3.7 MeV protons with a lithium-drifted silicon detector, obtained linearity of response from induced x-ray emission for thicknesses in the range 10^{-8} to 10^{-3} $kg\,m^{-2}$ for elements with $Z > 15$. They recommend proton scattering methods for lighter elements, since here the induced x-ray emission is of very low energy and is too difficult to measure accurately. A comparison of heavy-ion and optical methods for the measurement of thin metallic foils has been carried out by Muskalla *et al* (1981). They have compared measurements on unsupported aluminium, nickel, silver and gold films in the thickness range 2×10^{-4} to 5×10^{-3} $kg\,m^{-2}$ using multiple beam interferometry, $^4He^+$ absorption and $^4He^+$ backscatter methods. The accuracies are analysed and they conclude that the Rutherford backscattering integral method is to be preferred since, in addition to having an accuracy of 3%, it gives information about film contamination. The other nuclear methods described give comparable accuracy only for thicknesses greater than 10^{-3} $kg\,m^{-2}$ and require a precise knowledge of stopping power data.

4.2. Beta particle instrumentation

Most beta particles emitted from radioisotopes have a wide spectral range and although a few mono-energetic sources are available as a result of Auger transitions and internal conversion processes, most instrumentation has been based upon the wide-range energy sources.

The stopping of beta particles in matter is very complex involving several different mechanisms, and scattering processes have an important role in the interactions, so that the number of particles transmitted through an absorber cannot theoretically be related in a simple way to the absorber thickness or density. However, most response curves show that over a fairly wide range of absorber thickness $\ln R$ decreases linearly with thickness. This is true for most sources despite fairly wide variations in spectral form, so that the transmitted intensity I can be expressed in terms of the incident intensity I_0 by the relation $I = I_0 \exp(-\mu x)$ for much of the range. Moreover if thickness is plotted in terms of mass per unit area the value of μ is not strongly dependent upon the nature of the absorbing material. There is a small increase in μ with atomic number which is sufficient to introduce inaccuracies if variations in absorber composition occur which involve wide variations in atomic number. Transmission gauges are widely used in the paper industry and sensitivity to atomic constituents is important in this context since water content and the percentage of clays and titanium oxide in paper can vary considerably. For this reason it is often necessary to use these thickness gauges in conjunction with other forms of instrumentation. In the plastics industry where the overall atomic constituents are more predictable this problem is not serious.

The accuracy attainable with beta transmission gauges is very dependent upon maintaining fixed geometry since any slight variation in the relative positions of source, absorber and detector can produce variations in the scattered radiation received by the detector. The value of μ in $m^2\,kg^{-1}$ varies with the maximum energy E of the beta particles according to the empirical relation $\mu \propto E^{-1.14}$ so the choice of source for the measurement of any particular range of thickness measurement is important. A useful analysis of the optimum conditions for a thickness uniformity gauge is presented by Liljestrand et al (1976) covering thickness measurements up to $8\,kg\,m^{-2}$. For these measurements NE102 plastic scintillator was used as the detector. The importance is stressed of using the optimum value of μx for maximum sensitivity and statistical accuracy, the range of beta sources used having maximum energies ranging from $0.066\,MeV$ to $3.54\,MeV$ to correspond with thickness measurements from 3×10^{-3} to $18.6\,kg\,m^{-2}$ respectively. It was found necessary to discriminate against beta-induced x-rays in the absorber when thicknesses greater than $0.1\,kg\,m^{-2}$ were being measured, and for thicknesses greater than $8\,kg\,m^{-2}$ gamma attenuation methods are recommended due to the magnitude of beta scattering problems.

An on-line variation of the beta transmission gauge for which improved sensitivity is claimed is described by Pankratov et al (1976). This is used in the cellulose and paper industry in the Ukraine. Beta particles which have been transmitted through an absorber are deflected magnetically through 180° to pass back through the absorber again before entering the detector. An additional advantage of this procedure is that access to only one side of the absorber is required.

The backscattering of beta particles, which occurs as a result of elastic Coulomb interactions with nuclei, has been used for about 20 years for the measurement of sheet thickness and coating thicknesses of high atomic number materials on backing materials of low atomic number. It can also be used for the analysis of two-component materials and has the advantage of requiring access to only one side of the material under test. Backscattering of electrons is highly dependent upon the nature of the scattering material, the cross section for scatter increasing with the square of the atomic number. Since the particles are scattered from finite depths within the surface layer they will undergo energy losses before they emerge to enter the detector. The detector response will obviously be very dependent upon the geometrical arrangement and the beta particle energy as well as upon the atomic number of the scattering material, but it will be of the form

$$R = R_s(1 - \exp(-\mu x)) + R_0$$

where R_s is the maximum response for infinite thickness of the surface material and R_0 is the response for zero thickness, i.e. from the backing material and surroundings. $R_s = f_s R_0$ where f_s is known as the backscatter factor. In order to attain optimum sensitivity and accuracy conditions the energy of the source selected must increase with the thickness of the coating to be measured. With backscatter gauges the thicknesses which can be most accurately measured are smaller than with transmission gauges. Latter (1975) discusses the factors involved in controlling the sensitivity and reliability of beta backscatter measurements. He lists the differences in atomic number of coating and substrate which are necessary for 4% accuracy in the measurement of coating thickness over various ranges of atomic number, and explains the features of major importance in the geometrical design of the instrument. The test material must be in direct contact with a carefully designed aperture for consistency in measurements and due regard must be paid to all possible variables in the system if the output is to be obtained directly in terms of thickness. An 'in-process' application of the backscatter technique using ^{85}Kr beta particles is described by Traber (1974).

One comment frequently made about beta backscatter measurements is that very high instrumental stability is required for accurate values especially when the atomic numbers of the coating and substrate are not very different (Cameron 1972). Improvements in detection methods could give more reliable performance and encourage the use of beta backscatter methods in certain applications.

The excitation of x-rays in very thin films by beta particles can be used for the measurement of their thickness. Excitation can also be achieved by electrically generated electron beams. The two techniques are discussed by Timsit (1977) who compares the use of a ^{63}Ni beta source emitting particles of maximum energy 0.067 MeV with a 25 keV generated electron beam. The material examined consisted of an aluminium film on an iron substrate, the characteristic K x-ray lines being excited in both cases. Detection was by means of a cooled semiconductor energy analyser system which gave the relative intensity response of the two x-ray lines from the film and substrate. The ratio I_{Al}/I_{Fe} increased linearly with the aluminium thickness up to $4 \times 10^{-7}\,m$ when the ^{63}Ni source with its broad energy spectrum of beta particles was used, whereas the mono-energetic electron beam source gave a nonlinear response. This indicated the importance of the spectral composition as well as the energy of the incident particles in such measurements and their influence is not necessarily easy to predict. Calibration of the evaporated aluminium films was carried out with a quartz microbalance, and the estimated uncertainty in the thickness measurements was on average $\pm 7 \times 10^{-9}\,m$.

Beta-induced excitation of x-rays has been used by Jones et al (1972) for the assessment of particle size and distribution in ore. This method was particularly applied to gold particles and involved computer-controlled automatic scanning and analysis of test specimens.

4.3. Gamma instrumentation

Gamma radiation is the most penetrating of the ionising radiations and has consequently been well used in applications involving materials of high atomic number in dense layers. It is still probably the most widely used type of radiation in industry.

The absorption of mono-energetic gamma radiation by material of uniform density is strictly exponential, and since radioisotopes emit gamma rays of discrete energies the response in terms of count rate from a collimated source can very easily be related to the thickness of an absorber interposed between the source and detector. Some sources emit photons of more than one discrete energy and the fractional intensity transmitted for an absorber of thickness x is then given by

$$I_x/I_0 = \sum_{i=1}^{i=n} P_i \exp(-\mu_i x)$$

where P_i is the proportional intensity of radiation of energy i which has an absorption coefficient μ_i in the absorber. If the values of μ_i are appreciably different for different energy components the softer radiation must be filtered out or due allowance made for it in the calibration. The most sensitive type of detector for gamma radiation is the scintillation detector, the scintillator being chosen to have maximum absorption and light output for the gamma energy concerned. The most commonly used material is NaI(Tl). Scintillation detectors may also be used for energy measurement but the resolution is poor and cannot be compared with that attainable with cooled semiconductor detectors. However, the energy-dependent response of scintillation detectors enables the signal-to-noise ratio to be improved considerably by suitable pulse height discimination for simple count rate measurements.

Gamma transmission methods are commonly used for density measurements in fluids and slurries. In calculating the optimum gamma energy for such measurements due allowance must be made for pipe thickness and shape. Transmission gauges can also be used for the measurement of void fractions in pipes, e.g. in water/steam flow the void fraction is given by $\ln(I/I_w)/\ln(I_s/I_w)$ where I, I_w and I_s are the measured intensities after transmission through the same depth of the mixed fluids, water and steam respectively. With all transmission-type measurements, it is desirable to avoid the detection of scattered radiation, so collimation of the source and screening of the detector from all but the directly transmitted radiation is important. The application of gamma radiation to density measurements is described by Springer (1979) with particular reference to current instrumentation in South Africa.

Gamma transmission methods have been widely employed in the design of liquid level gauges. An interesting extension of this principle has been used for the assessment of an inter-zone layer in an extraction tower which is described by Elias et al (1974a). Details are given for the particular case of an oil extraction plant, but the mathematical model set up gives guidelines which could be adapted to predict the optimum conditions for other applications.

Gamma transmission methods are used to measure mass per unit length on loaded conveyor belts and, in conjunction with a tachometer, can measure mass transported per unit time. Boyce et al (1973) describe this application discussing the relative advantages of 'point' and 'line' sources and various types of detectors. When used for belts loaded with pellets, fines and rubble with optimum geometry the errors can be less than 1%, but they increase with particle size. These factors are also considered by Colijn (1975) with particular reference to the requirements of the metal industry. A further discussion of errors in this type of measurement is given by Hold et al (1973) and comparison is made with other forms of instrumentation. Elias et al (1980) analyse the accuracy and sensitivity of nuclear belt weighers over a wide variety of conditions. They discuss in detail certain design parameters which strongly influence accuracy and sensitivity and in particular show how these are affected by load values.

Instrumentation for the measurement and control of thickness by alpha transmission gauges in strip metal rolling is discussed by Perriam (1972) and also by Boyle (1972). In such applications the source-to-detector distance must be large and the response time less than 100 ms. Miyagawa (1973) describes gamma thickness instrumentation used in the Nippon steel industry in Japan and compares this with non-nuclear techniques.

Fishman et al (1981) describe a transmission gauge to assay the integral water content in soil. By burying a 2.3 mCi Cs source 0.53 m below the surface of the ground a water content of approximately $200\ \mathrm{kg\ m^{-3}}$ was measured with 2 to 3% resolving power.

Allen and Svarovsky (1972) describe the use of x-ray transmission for the detection of particle size. This is applied to the standard sedimentation technique with the usual optical examination of the sedimentation tank replaced by an isotope–detector transmission system. A 3 Ci Pr–Al isotope source was used in this case with a proportional counter as detector.

The sedimentation method for particle size measurement can also be used with tracer techniques (Gardner and Ely 1967). For tracer methods short half-life gamma emitters are required in order to reduce safety problems. This is in complete contrast to the normal requirement in nucleonic instrumentation which is for a source which will give an essentially constant emission over a long time period. Gamma emitters have been used as tracers to monitor flow patterns in industry and in civil engineering (e.g. Wurzel 1971), but alternative methods are used wherever possible due to safety hazards.

The cross section for scattering of high-energy gamma radiation by the Compton effect is proportional to $\rho Z/A$ where ρ is the density, Z the atomic number and A the atomic weight of the scattering material. Under conditions of fixed geometry one

115

might therefore expect the response of a detector shielded from the primary beam and therefore receiving only scattered radiation from the surrounding medium to be directly proportional to the density of the medium and not very sensitive to its chemical composition. However, the scattered radiation, being of longer wavelength than the primary radiation, is attenuated by photoelectric absorption which is strongly dependent upon the atomic number. Consequently the response of a gamma scatter gauge increases nonlinearly with the density of the surrounding medium until a maximum response is reached, and then falls off in an approximately exponential manner. There is also a marked difference in response for scattering materials of different atomic number. Ballard and Gardner (1965) have used a comprehensive mathematical model to predict gauge response to various materials and have applied this to soil density gauges. Their treatment is summarised by Gardner and Ely (1967). Gamma backscatter gauges are used in the coal industry as control probes in the steering system for coal-cutting machines. These measure the thickness of coal seams by utilising the difference in Compton scattering by coal and by the surrounding shale which have specific gravities of about 1.4 and 2.0 respectively. Cooper (1972) compares probes using ^{241}Am and ^{137}Cs sources and discusses problems of instrumentation and errors arising due to the arduous conditions of operation. Elias et al (1974b) have set up a computerised model to determine the response function of a gamma radiation dual gauge for density measurements. This utilises information concerning both the primary beam attenuation and the scattered component, and can therefore give more information regarding the physical characteristics of the interacting medium.

It has been common practice for many years to combine information obtained by different techniques to determine the density and composition of soil. The variation in natural gamma activity of different types and thicknesses of rock strata, gamma backscatter measurements and various neutral techniques have been used in combination, particularly in oil exploration (Cameron and Clayton 1971). Modern methods of processing such data can give reliable information very rapidly.

4.3.1. X-ray fluorescence. The inner-shell electrons in atoms readily absorb x-ray photons which have exactly the energy required to release them from the atom. Vacancies left in the atomic structure are then filled by other electrons, the surplus energy being emitted as low-energy x-rays with wavelengths characteristic of the particular type of atom. This process is known as x-ray fluorescence, and although it involves the use of ionising electromagnetic radiation the energies used are lower than in the gamma radiation measurements previously described and the measuring techniques and scope of use are therefore different.

X-ray fluorescence techniques have been widely used for the chemical analysis of surfaces and they can also be used with considerable advantage for the measurement of thickness of very thin layers.

The probability that an incident photon will produce a characteristic x-ray in a thickness element dx at a depth x below a surface which will be detected at the surface is

$$P \, dx = P_1 P_2 P_3 \, dx$$

where P_1 is the probability that an incident photon will not be absorbed in the initial thickness x, and P_2 is the probability that the photon will excite the required characteristic radiation. For P_2 to be a maximum the energy of the incident photons should be a little higher than the characteristic energy of emission. P_3 is the probability that an emitted photon will not be absorbed before reaching the detector. If μ_1 and μ_2 are the absorption coefficients of the surface material for the incident and emitted radiation respectively, the count rate registered by the detector due to the surface film will increase with its thickness in proportion to

$$(1 - \exp[-(\mu_1 + \mu_2)x]).$$

Gardner et al (1972b) discuss mathematical models for x-ray fluorescence gauges which analyse the factors governing instrumental response. Instrument design involves the selection of x-rays of suitable energy to excite maximum fluorescence, and this may be achieved either by a suitably chosen radio-isotope or by electrical generation with appropriate filtration or energy selection. Higher intensities are available from machines but these are more cumbersome, more expensive and less constant in output. Clayton et al (1973) stress the importance of using monochromatic radiation of the optimum energy in these measurements and recommend electrical generation in conjunction with an energy selector. On the other hand, Cameron (1972) stresses the advantage of isotopic sources in a discussion of typical instrument specification for on-line control. The choice would seem to depend upon a number of factors such as the availability of a suitable isotope for a particular application, working conditions, required accuracy, etc. Since the emitted x-rays are of low energy and easily absorbed the source–sample–detector geometry is very important and detection is preferably by a cooled lithium-drifted solid state detector which will give a linear response to the energy of the emitted x-rays, and spectral analysis of the output enables direct programming of the instrument to give the layer thickness. An assessment of the advantages and disadvantages of lithium-drifted solid state detectors for use in on-line applications of x-ray fluorescence measurements is given by Carr-Brion (1972). Wendt (1973) describes thin-film studies by x-ray fluorescence particularly in order to investigate the limits of detectability and measurement of thin films. Thickness measurement was found to be possible over five orders of magnitude under very clean substrate conditions. For very thin films the detector response was proportional to thickness so that calibration was particularly simple. The method gives no contamination or heating problems which are sometimes encountered in electron probe methods.

Kaushik et al (1980) used the technique to determine the

thickness of single, double and triple layered films on Mylar substrates. For this they used an annular 5 mCi source of ^{109}Cd with a lithium-drifted silicon detector. The minimum thickness detection limits were 70, 70 and 40 nm for Cu, Bi and Au layers respectively. They compare these results favourably with earlier measurements by Singh et al (1979) using an ^{241}Am source.

Rosner et al (1975) compare methods of measuring coating thickness based on the intensity ratio of various characteristic lines emitted by the substrate, the ratio of various characteristic lines emitted by the coating and the ratio of radiation emitted by the base to that emitted by the coating. They conclude that the last of these methods gives the best resolving power by a factor between three and six for Au coating on Ag but that the other ratio methods may be useful in specific cases, e.g. where the coating is of a light element and the emitted x-rays are therefore difficult to detect, then the first method may be advantageous.

Rötzer and Markowitsch (1972) describe the instrumentation used in the application of x-ray-induced fluorescence to measure the thickness of layers of varnish (10 to 20 μm) on the inside and outside of aluminium tubes. Since the inner coating is of a very low atomic number material the attenuation of the x-rays excited in the aluminium is used as a measure of the coating thickness. The varnish used for the outer layer, however, contains titanium and the x-ray intensity emitted by this is found to give a useful measure of the outer varnish thickness. An unsealed proportional flow counter is used for detection since the x-rays are of very low energy. The precision quoted is $\pm 0.35\,\mu$m and $\pm 0.5\,\mu$m for measurement of the internal and external coatings respectively.

Luzzi et al (1980) describe a simple, quick and cheap application of x-ray-induced fluorescence for the industrial measurement of metallic layers of up to 10^5 nm thickness with an accuracy which may be sufficient for many purposes. They use a 30 mCi ^{241}Cm source with a proportional gas counter as detector, and give estimates of the minimum detectable thickness for various coating/substrate combinations in a 200 s counting time.

X-ray fluorescence is now acclaimed by many as the most accurate and reliable technique for measuring thin metal coatings (Donhoffer and Beswick 1973, Cameron 1972, Gruber 1972). Notea and Segal (1974b) discuss the effect on the response function of these thickness gauges due to secondary x-rays produced in the base which also initiate excitation in the coating. This secondary effect can completely change the dependence relation between the sensitivity and the coating thickness, and is indicative of anomalies which can arise due to oversimplification in the analysis of instrumental factors in x-ray fluorescence and may therefore lead to the adoption of conditions which are not optimal.

5. Conclusions

The use of ionising radiation for the measurement of physical parameters is now very widespread and it has been possible to mention only briefly the main techniques of measurement. A British Standards Institution guide to ionisation radiation thickness meters for materials in the form of sheets, coatings or laminates has been published (*British Standards Institution* 1980) which discusses error limits and general performance tests.

The main improvements in instrumentation which can be expected are in the use of improved detectors, the optimisation of design for specific applications and the introduction of dual gauging techniques in situations where composition as well as thickness or density may be varying. This can involve the use of mixed energy radiation or mixed types of radiation, e.g. photons and neutrons (Dunn and Gardner 1975). Further application of computerised systems for data processing should make the analysis of more complex physical and chemical parameters feasible.

Acknowledgments

The author would like to thank Mr S Shaw and his colleagues at The City University library for their help in the information search which was essential in the preparation of this article, Mr G A Aldous for helpful discussion and Professor L Finkelstein for his encouragement.

References

Allen T and Svarovsky L 1972 Particle size determination by light and by x-ray attenuation
Industrial Measurement and Control by Radiation Techniques IEE Pub **84** pp 7–15

Ballard L F and Gardner R P 1965 Density and moisture content measurements by nuclear methods
NCHRP Report 14 Highway Research Board, Nat. Acad. Sci-Nat. Res. Council

Boyce I S, Cameron J F and Piper D 1973 Nucleonic conveyor-belt weighers for the basic metals industry
Nuclear Techniques in the Basic Metal Industries, Proc. of Symp. (Helsinki) IAEA **159** pp 155–64

Boyle P W 1972 The application of radioisotopes to the measurement and control of strip thickness in the steel industry
Industrial Measurement and Control by Radiation Techniques IEE Pub **84** pp 232–7

British Standards Institution 1980 Guide to ionisation radiation thickness meters for materials in the form of sheets, coatings or laminates
BS 5868

Cameron J F 1972 On-line measurement and control of continuous tin and zinc coating lines by radioisotope x-ray fluorescence techniques
Industrial Measurement and Control by Radiation Techniques IEE Pub **84** pp 72–7

Cameron J F and Clayton C G 1971 *Radioisotope Instruments* (Oxford: Pergamon)

Carr-Brion K G 1972 On-stream application of semiconductor x-ray detectors
Industrial Measurement and Control by Radiation Techniques IEE Pub **84** pp 103–7

Clayton C G, Packer T W and Fisher J C 1973 Primary energy selection in non-dispersive x-ray fluorescence spectrometry for alloy analysis and coating thickness measurement
Nuclear Techniques in the Basic Metal Industries, Proc. of Symp. (Helsinki) IAEA **159** pp 319–54

Colijn H 1975 Nuclear weigh scales weighing and proportioning of bulk solids
Trans. Tech. Publ. 286–304

Cooper L R 1972 Gamma-ray backscatter gauges for measuring thickness on mechanical coalfaces
Industrial Measurement and Control by Radiation Techniques IEE Pub **84** pp 89–93

Donhoffer D K and Beswick C K 1973 Data processing and control systems in metal coating operations
Nuclear Techniques in the Basic Metal Industries, Proc. of Symp. (Helsinki) IAEA **159** pp 299–317

Dunn W L and Gardner R P 1975 Development of mathematical models and the dual gauge principle for surface-type neutron moisture content gauges
Nuclear Engineering and Design **32** 438–48

Dyson N A 1973 *X-rays in Atomic and Nuclear Physics* (London: Longman)

Eichholz G G and Poston J W 1979 *Principles of Nuclear Radiation Detection* (Ann Arbor: Ann Arbor Publishers)

Elias E, Pieters W and Yom-Tov Z 1980 Accuracy and performance analysis of a nuclear belt weigher
Nuclear Instrum. Methods **178** 109–15

Elias E, Segal Y and Notea A 1974a Gamma gauge for the control of interzone layer in an extraction tower
Nucl. Technol. **21** 57–66

Elias E, Segal Y and Notea A 1974b Dual gauging utilising penetrating and scattered photons
Trans. Am. Nucl. Soc. **19** 79–80

Evans R D 1968 X-ray and γ-ray interactions *Radiation Dosimetry* Vol. I ed. F H Attix and W C Roesch (New York: Academic Press) pp 93–153

Fishman A, Notea A and Segal Y 1981 Gamma transmission gauge for assay of integral water content in soil
Nuclear Instrum. Methods **184** 571–7

Gardner R P and Ely R L 1967 *Radioisotope Measurement Applications in Engineering* (New York: Reinhold)

Gardner R P, Verghese K and Betel D 1972b Simple mathematical models for non-dispersive x-ray fluorescence gauges
Trans. Am. Nucl. Soc. **15** 708–9

Gardner R P, Verghese K and Cehn J I 1972a Mathematical model for the measurement of gas properties with alpha particle gauges
Nucl. Technol. **16** 418–29

Gruber U 1972 Measuring the thickness of the zinc coating on continuously galvanized steel sheet
Industrial Measurement and Control by Radiation Techniques IEE Pub **84** pp 136–42

Hold A C, Morgan D W and Williams K F 1973 Application of gamma-ray absorption to the determination of flow of materials on conveyor systems
Nuclear Techniques in the Basic Metal Industries, Proc. of Symp. (Helsinki) IAEA **159** pp 165–82

International Commission on Radiological Protection 1977 ICRP publication **26** (Oxford: Pergamon)

Johannson T B, van Grieken R E, Nelson J W and Winchester J W 1975 Elemental trace analysis of small samples by proton induced x-ray emission
Anal. Chem. **47** 855–60

Johns T E 1982 The purpose and practice of radiation monitoring
IEE Proc. **129** pt A no 2 1982 81–105

Jones M P, Beaver C H J and Shaw J L 1972 The measurement of stereological parameters for the mineral industry by a computer controlled electron probe x-ray microanalyser
Industrial Measurement and Control by Radiation Techniques IEE Pub **84** pp 1–6

Kaushik D K, Singh S P, Bhan C, Chattopadhyaya S K and Nath N 1980 X-ray fluorescence to determine the thickness of single, double and triple-layered films of Cu, Bi and Au on Mylar substrates
Thin Solid Films **67** 353–6

Kerkow H, Kryesch G and Stolle R 1979 Measurement of the thickness of thin layers by proton induced x-ray emission
Thin Solid Films **62** 73–81

Kowalski E 1970 *Nuclear Electronics* (Berlin: Springer)

Kurstedt H A 1976 Heuristic approaches in the design and use of nuclear instrumentation
Proc. 22nd Int. Instrum. Symp., San Diego, California pp 595–602

Latter T D T 1975 Measuring coating thickness by beta backscatter techniques
Br. J. Non-destructive Testing **17** 145–52

Liljestrand R, Blanpied G and Hoffman G W 1976 Target thickness uniformity gauge
Nucl. Instrum. Methods **138** 471–7

Luzzi G, Mazzel A, Neri A, Salmi M and Schirripa Spagnolo G 1980 Measurement of thin film thickness by means of a simple, non-destructive radioisotope technique
Thin Solid Films **67** 347–51

Matsumoto S and Kobayashi H 1981 Humidity measurements using a double ionisation chamber with alpha rays
Nucl. Instrum. Methods **184** 603–7

Miyagawa K 1973 Application of measurement systems using radioisotopes in the Nippon Steel Corporation
Nuclear Techniques in the Basic Metal Industries, Proc. of Symp. (Helsinki) IAEA **159** pp 287–97

Muskalla K, Pfeiffer Th and Moller W 1981 A comparison of optical and nuclear methods for the measurement of the thickness of thin metal foils
Thin Solid Films **76** 259–67

Nicholson P W 1974 *Nuclear Electronics* (New York: Wiley–Interscience)

Notea A and Segal Y 1974a A general approach to the design of radiation gauges
Nucl. Technol. **24** 73–80

Notea A and Segal Y 1974b Sensitivity variations due to secondary excitations in x-ray fluorescence coating gauges
Trans. Am. Nucl. Soc. **19** 78–9

Pankratov V I, Prilipko I T, Gomberg M L, Zhikharev V V and Adamenko A A 1976 Radioisotope thickness gauge based on a stream of β particles deflected by a magnetic field
Measurement Techniques (USA) **19** 824–6

Perriam L T 1972 X-ray and nuclear techniques for thickness measurement in strip metal rolling
Industrial Measurement and Control by Radiation Techniques IEE Pub **84** pp 153–8

Radiation Sources Catalogue 1981 (Amersham: Amersham International PLC)

Robinson F N H 1974 *Noise and Fluctuations in Electronic Devices and Circuits* (Oxford: Clarendon)

Rosner B, Notea A and Segal Y 1975 Thickness gauging by x-ray fluorescence
Thin Solid Films **27** 141–7

Rötzer H and Markowitsch P 1972 Non-destructive thickness measurement of inside and outside coatings of aluminium tubes
Industrial Measurement and Control by Radiation Techniques IEE Pub **84** pp 83–8

Segal Y, Notea A and Segal E 1977 Systematic evaluation of non-destructive testing methods
Research Techniques in Non-destructive Testing vol. III ed. R. S. Sharpe (New York: Academic Press) pp 293–322

Singh S P, Kaushik D K, Chattopadhyaya S K and Nath N 1979 Thickness measurements of single and composite thin metal films using the x-ray fluorescence technique
Thin Solid Films **59** 51–5

Springer E K 1979 Density measurement using gamma radiation – theory and application
South African Mech. Eng. **29** 443–52

Tait W H 1980 *Radiation Detection* (Sevenoaks: Butterworth)

Taylor D 1971 Radioisotope instrumentation
Proc. IEE **118**(9R) 1090–106

Thomas J P and Cachard A 1978 *Material Characterisation Using Ion Beams* (New York: Plenum)

Timsit R S 1977 A ^{63}Ni radioactive source for determination of thickness of supported metallic films
Thin Solid Films **47** 323–6

Traber K 1974 Continuous in-process measurement of zinc layer thickness
Fertigungstechnik und Betrieb **24** 434

Washtell C C H and Hewitt S G 1965 *Nucleonic Instrumentation* (London: Newnes)

Wendt M 1973 Thin film studies by x-ray fluorescence
Phys. Status Solidi a **16** 115–20

Williams J 1979 Tips on nuclear gauging for level and density measurements
Instrum. and Control Systems (USA) **52** 47–51

Wurzel P 1971 Radioisotopes in some aspects of civil engineering
The Rhodesian Engineer **9** 949–53

Chapter 10

Measurement for and by pattern recognition

C J D M Verhagen
Pattern Recognition Group, Department of Applied Physics,
Delft University of Technology, Lorentzweg 1, 2628 CJ Delft,
The Netherlands

Abstract. The relations between measurement and pattern recognition are discussed. Pattern recognition requires measurement as its first step and the purpose of many measurements is a classification – a recognition of the class to which the unknown input situation belongs. Some general aspects of pattern recognition are discussed, especially the notion of variability in the patterns belonging to one class. A relation between pattern recognition and nominal measurement is indicated. Two methods often encountered in solving pattern recognition problems are briefly described. Finally some remarks concerning sensors for pattern recognition problems are made. Two survey papers on pattern recognition containing many references to the literature are mentioned.

1. General introduction

Pattern recognition in our context always means 'automatic' pattern recognition.

The title of this paper indicates a double relation between measurement and pattern recognition.

In the first place the title refers to the fact that in many pattern recognition problems one has to start from data obtained by physical measurement; patterns (classes) have to be recognised in the data. For example, the reflected light from many points on an illuminated surface has to be measured before pattern recognition methods can be applied to recognise the patterns of the characters on the surface.

Pattern recognition problems, however, do not only start from data obtained through physical measurement, but also from, for example, economic data. With a broad definition of 'measurement' (for instance, measurement means mapping real world situations to the domain of numbers) this type of data may originate from measurement too. Thus it makes sense to state that pattern recognition problems start with some kind of measurement.

A comment might be useful in order to avoid misunderstanding. If 'measurement' always includes the actual mapping to numbers (as the given definition of measurement suggests), then measurement itself is not essential as a first step to pattern recognition. It is sufficient for sensors to produce signals related to the unknown (in the example mentioned, the voltages from a sensor scanning the reflected light from the surface with the characters) and for recognition to follow from processing these signals, without having to go to numbers during the first step.

A corresponding situation exists, for instance, in control engineering; here the signals from sensors may immediately be compared with reference signals in order to determine a control action by feedback. In such situations it is also customary to state that measurement is necessary to enable control, though no numerical values are present.

A similar situation exists when using an indicating measuring instrument but without an observer to map the position of a pointer to a number. In everyday speech the notion 'measurement', however, is used in this case too though no actual numbers are produced.

With this comment in mind one may say that pattern recognition involves measurement as its first step. However, it would be more precise to state that the first step requires a sensor (again with a broad definition of 'sensor').

Sometimes 'normal' sensors are used in pattern recognition, e.g. microphones to measure sound pressures for speech recognition or photocells to measure reflected light. In many cases, however, sensors have to be adapted or new ones developed in order to produce signals suitable for pattern recognition. In this respect the pattern recognition field has stimulated sensor development and thus contributed to the measurement field. In § 5 some remarks are made about this development.

In the second place the title of this paper suggests that measurement may 'need' pattern recognition. Classically, the purpose of measurement has been to determine the numerical value of some quantity to verify a physical hypothesis. With increasing frequency, however, measured data are processed, combined and compared with other data in order to obtain a specific conclusion. In some situations this conclusion may be a simple binary (yes/no) decision. Examples might include quality control in industry (good versus bad) or medical testing (within normal limits or not). In other areas the conclusion may be subdivided into many classes. Such multiclass decision making might include medical diagnosis (which disease?) or particle physics (what type of particle made this track?). In these examples the purpose of the measuring process is not to produce numerical values for the variables but to generate meaningful interpretations of the variables. Such a classification procedure may be interpreted as a recognition of the class to which the input belongs. In complex situations where pattern recognition methods have to be applied, this is a pattern recognition problem.

In the cases considered here, pattern recognition is an essential step in the measuring procedure. Thus it may be argued that pattern recognition is very closely related to measurement, that it is a consequence of measurement, or that it is a part of the measuring process.

Finally it may be mentioned that in pattern recognition, in addition to a classification into say, good or bad, sometimes a numerical estimation can be given about how good or how bad a situation is, or membership values or probabilities can be added to the classification results (see also 'fuzzy' approaches in the next section). It follows again that measurement and pattern recognition are strongly interrelated.

Measurement and pattern recognition have usually been treated as two different disciplines, each having its own methods, traditions, journals and conferences. In many situations, however, a multidisciplinary or interdisciplinary approach would be useful and should be stimulated. A measuring part (sensor) well matched to the pattern recognition algorithm is most desirable. The best situation is when the data, obtained from the sensor, are in a form (with respect to format, noise, disturbances) that can easily be treated by the pattern recognition algorithm. To achieve this a close cooperation between experts from both the measurement and the pattern recognition field is required.

In education it is advisable to have a course on pattern recognition follow one on measurement, especially for those departments in which (as a rule) operational conclusions have to be drawn from measurements; this is certainly true in engineering education. It may be interesting to mention that several research groups engaged in measurement have gradually shifted their main interest to pattern recognition problems. The author's group is an example.

In addition to a close relation between pattern recognition and measurement, pattern recognition is also strongly allied with such fields as statistics, information theory, computer science, etc. The interplay and diffusion of ideas across these fields have helped to strengthen pattern recognition methodology. It has also been observed that some groups working in these fields have shifted their attention to pattern recognition.

2. General aspects of pattern recognition

Only a few general remarks concerning pattern recognition are possible within the scope of this paper. The survey paper of Verhagen *et al* (1980) gives a more extensive treatment of this subject. A second survey (Verhagen *et al* 1982) deals more briefly with the general aspects of pattern recognition, but pays more attention to its relation to measurement and to measuring systems appropriate for pattern recognition. The present paper contains in a nutshell many ideas from this second survey.

An analysis of pattern recognition problems reveals that complex input data containing one or more patterns are to be mapped at the output of the system into the class indications of the patterns. Hence the output is quite simple.

It is not necessary that a certain pattern be mapped into only one class; in 'fuzzy' approaches a mapping to a number of classes, each with a certain membership value or probability, may be appropriate.

Some of the main methods for the mapping are briefly discussed in § 4.

Some classical examples of pattern recognition problems are the recognition of characters, speech, chromosomes, blood cells, etc.

Note that, apart from 'fuzzy' approaches, a single class may represent many, quite different, input situations. This requires a 'many-to-one' mapping. An obvious example is that all the different ways the letter 'A' (or 'a') can be written, typed or printed belong to the class 'A', so that all inputs produced by a sensor from specimens of all 'A's or 'a's have to be mapped to the same output indication of class 'A'.

The origin and the contents of the variability in the patterns belonging to one class can be very different. The variability that may be present in 'cultural' patterns (such as those determined by characters, cars, houses, etc) depend on – changeable – human convention; the variability of 'natural' patterns belonging to biological objects (such as chromosomes, blood cells, electrocardiograms, etc) is determined by nature and is often strongly influenced by the way they are prepared. In all cases the experimental situation (e.g. illumination, direction of viewing) and the sensor properties (e.g. noise, distortion) may produce additional variability. As with measurements it is desirable to keep this additional variability small in relation to the 'essential' variability of the patterns.

It should be noted that class definition depends usually on human deliberation or tradition; this determines what types of different input situations have to be taken together into a certain class. Class definition may also depend on the aggregation level wanted. One may, for example, be interested in recognising the presence of animals or vehicles in a scene, or in recognising a specific type of animal (dog) or vehicle (tank), or even in identifying an individual specimen. After choosing the set of input situations that has to form one class (with a specific pattern), this set constitutes a so-called 'learning set'. The pattern recognition system has to recognise a specimen of this learning set but also all 'equivalent' specimens (a more formal discussion of the concept of 'equivalency' is given in the survey papers mentioned above). An exception to this procedure is the pattern recognition problem 'without a teacher', that is, without a learning set with given class labels. Here, the system must 'teach itself' by means of a technique called 'cluster analysis' what types of equivalent situations or patterns may exist in the input data.

Finally it may be noted that it is customary to use the notion 'pattern recognition' only for 'complex' situations where the effect of variability is marked. 'Simple' classification problems (for instance, whether a measured temperature is within a safe temperature interval or not) are not denoted as 'pattern recognition'. As usual in these situations there is only a 'fuzzy' boundary between 'simple' and 'complex'.

3. Pattern recognition and nominal measurement

It may be useful to indicate a resemblance between pattern recognition and a specific type of measurement, the 'nominal' measurement. Nominal measurement is often considered as rather elementary, introduced in the very beginning of textbooks on measurement to explain the measurement process. Essentially it means comparing a certain aspect of unknown items with a standard set of reference items. A classical example is the comparison of colours with a standard set of, say, the colours of next year's cars. The unknown colour receives the name of that item of the reference set that (nearly) equals the unknown. The procedure seems quite elementary as one only needs a comparison with respect to (near) equality to the reference items; no 'order', no mapping to a scale with ordered numbers is necessary. Instead of a name, of course, a symbol or even a number may be chosen, but the properties of numbers do not apply.

One may describe nominal measurement as a classification procedure where the input data are mapped into a class of the reference set. If no (sufficient) equality with any item of that set exists a mapping to the class 'reject' can be made. The resemblance to pattern recognition seems obvious.

One, however, has to bear in mind the following points. In explaining nominal measurements the comparison is often made by human observers, who look at the unknown item and at the reference set (as in the case of colours). In (automatic) pattern recognition this has to be done by means of 'measurements' and often sophisticated methods to 'compare' data.

In pattern recognition a well-defined standard set of reference items does not exist. An elementary-school teacher may try to teach ideal shapes of numbers and letters, but in practice a great variety of shapes result and are accepted. These shapes are not well-defined but are given by cultural tradition or, in the case of biological items, by nature. Now instead of a standard set of reference items, learning sets (see previous section) have to be used for all the classes (the patterns) to be recognised.

Adding the concept 'near' to the description of nominal measurement as given before allows some deviation from 'equality'. This indicates some similarity to the concept of variability as used in pattern recognition problems. If in the description of nominal measurement one were to substitute 'a comparison as to equivalency' (with the concept 'equivalency' as indicated in the previous section) instead of 'a comparison with respect to (near) equality', the resemblance between nominal measurement and pattern recognition would be closer. In that case, however, nominal measurement is used in a broader and more sophisticated way than is usually implied in textbooks.

4. Pattern recognition methods

The survey papers mentioned before and their many references give an overall picture of the methods that are often used in pattern recognition. Because of the very different types of patterns to be recognised in one, two, three and even more dimensions, and the very different types of variabilities that may be present, it is not surprising that a great variety of pattern recognition methodologies exist, including the occasional *ad hoc* solutions.

This section outlines two often encountered methods. Both use as a first operation the determination of 'features', but the two methods differ greatly as to the nature of the features. Features are produced by processing the input sensor data which greatly reduces the data sets; the feature values are expected to be characteristic for the patterns to be found. Data reduction by means of features is necessary in almost all pattern recognition methods because of the fact that the input sensor data are usually very complex and contain a lot of irrelevant and redundant data. In two-dimensional images, which are very common in pattern recognition problems, it is customary to use a square grid of *pic*ture *el*ements (pixels) with, for example, 512×512 pixels and 8 bits to represent each pixel, resulting in more than 2×10^6 bits. This has to be reduced at the output to say 100 classes, or less than 7 bits. This reduction is performed in (at least) two steps: an initial reduction into features is followed by a decision process based upon the feature data to find the classes involved.

The choice of good features usually requires a great deal of *a priori* knowledge about the patterns; if this knowledge is not present, trial and error together with, for example, statistical evaluation may be necessary, demanding a large learning set.

The first of the often encountered methods is the statistical approach. It employs as features the values of a number of quantities (e.g. the height and length dimensions of cars and lorries, the coordinates of characteristic points in curves, the size and position of bands in chromosomes). These values may be determined by direct measurement, but intensive processing of sensor data is frequently necessary.

If k features are chosen, their k values may be used to indicate a point in a k-dimensional feature space. The variability as discussed in § 2 causes a spread of the sample points of the learning set for each pattern. This results in a 'cluster' of points per pattern. Good features together with quite different types of patterns will result in a small spread for each of the pattern clusters and relatively long distances between the clusters of different patterns. If the clusters do not overlap discriminant planes can be constructed that completely separate the clusters. This permits correct classification decisions about unknown specimens with their feature space points near the clusters. Usually, however, the great spread in the feature values and the small distances between clusters produce overlapping clusters. In this case decision algorithms have been developed to construct some kind of 'optimal' discriminant planes, which depend upon the statistical knowledge of the probability densities of the clusters.

The second method is the linguistic approach. It is based on a decomposition of a pattern into elementary parts or 'primitives' that together compose the pattern. Straight and curved line segments, for instance, may compose a line figure. The primitives, their positions, and their relations to each other are used as features of the patterns involved. The relations between the primitives can sometimes be expressed by a kind of grammar, similar to a linguistic grammar which describes the relations between the elementary parts of a sentence. Many different types of grammar have been developed for pattern recognition purposes, including stochastic grammars allowing variability.

Recognition now requires the determination of the primitives, their positions and relations, followed by a decision process called 'parsing' to determine which pattern grammar matches the situation. A grammar may follow from *a priori* knowledge or may be inferred from a learning set. Again many decision algorithms have been developed, depending on the type of grammar and the available *a priori* knowledge.

These two methods, statistical and linguistic, are by no means fully developed and further work needs to be done. In addition, as the range of pattern recognition applications grows, insight into the two methods will grow too. Even so, occasional *ad hoc* solutions may be required for special problems.

5. Sensors and processors for pattern recognition

The survey of 1982 (Verhagen *et al* 1982) treats a number of pattern recognition sensors and refers to the appropriate literature. Here, only an outline is offered. We shall restrict the discussion mainly to two- and three-dimensional optical image sensors (or scanners) where a great deal of progress has been made.

As most pattern recognition algorithms use a digital representation of quantities, a 'normal' analogue sensor has to be followed by an analogue-to-digital converter. In the case of images with very many pixels, fast converters are often required. TV cameras which scan two interleaved frames 25 times a second with 625 lines in the two frames have a usable line time of about $50\,\mu s$. To obtain in that time 512 points the sampling frequency must be as high as 10^7 Hz. As this speed goes beyond the input possibilities of most general purpose computers, special interface memories and often special processors are necessary.

In addition to optical images, attention has been given to images originating from nuclear, x-ray and ultrasonic sources and also from spin imaging in the field of nuclear magnetic resonance. From these sources data about the internal structure of opaque objects can be obtained which are beyond the capabilities of optical images. The results may be sensed and digitised immediately or an intermediate photographic step may be required.

Some of the optical scanners for pattern recognition are closely related to traditional scanners; they include digitised versions of moving-stage microscopes, drum scanners, and TV, flying-spot or photo-array scanners.

Scanners for remote sensing of the earth from aircraft or satellites produce data in several narrow wavelength bands (also in the infrared) from which geological, agricultural or geographical situations may be recognised.

Three-dimensional information can be obtained from one two-dimensional image by stereopsis methods or from a number of two-dimensional projections in different directions by tomographic methods. For robot systems, which handle a limited set of three-dimensional objects, special illumination strategies are made possible by using planes of light shifted parallel to or rotated around an axis; the shapes and positions of the objects can be determined from the illuminated set of lines on the surface of the objects. Measuring light flight times of laser beams reflected on the surface of an object also produces data about the third dimension.

A sensor will often introduce distortions originating from, for example, its point-spread-function, its noise or from its motion with respect to the object. A restoration of the sensor output by preprocessing may thus be required before further pattern recognition processing takes place. Examples of such preprocessing might include inverse filtering, linear least-squares filtering and edge-preserving filtering. To make feature determination easier an intentional distortion of the image (called 'image enhancement') may be useful; lateral high-frequency filtering, for example, can enhance the edges of contours which are important for shape recognition.

Conventional computers are often too slow for such preprocessing, and sometimes also for feature extraction from images, for effective applications in practice. A special-purpose (pre)processor can sometimes be combined with an appropriate sensor. Several solutions to increase speed are possible: special software, faster hardware systems, and especially some form of parallelism. Parallelism may be realised in pipelined processors, multiprocessors, array processors, etc. Certain types of processing can even be achieved at TV speed, allowing real-time processing. With somewhat slower processing, for instance of the order of one second, interactive operation with a human operator, who can see on a display almost immediately the effect of a certain operation, is attractive.

In the design and learning stage of a pattern recognition system general purpose computers may be useful because many alternatives have to be compared and speed is not essential. For application in practice, however, more specialised and faster systems are usually necessary. The option of combining 'smart' sensors with (pre)processors (made possible by the microprocessor) looks very promising for the construction of such fast and not too expensive components for practical systems.

These developments in sensors combined with (pre)processing, together with developments in feature extraction and decision algorithms, make it reasonable to expect that we shall see an increase in the application of pattern recognition methodologies and instrumentation in practice.

Acknowledgments

The author wishes to thank all members of the Pattern Recognition Group at the Delft University of Technology for their contributions to the ideas developed in this paper. He acknowledges with pleasure the critical comments on this paper by Professor A Choudry, Dr R P W Duin, Dr F C A Groen, Dr P W Verbeek and especially by Professor Dr I T Young. Mrs S Massotty advised on the English language.

References

Verhagen C J D M, Duin R P W, Gerritsen F A, Groen F C A, Joosten J C and Verbeek P W 1982 *Handbook of Fundamentals of Measurement Systems* Vol. 1 *Theoretical fundamentals* (ed. P Sydenham) (Chichester: Wiley) Chap. 7

Verhagen C J D M, Duin R P W, Groen F C A, Joosten J C and Verbeek P W 1980 *Rep. Progr. Phys.* **43** 785–831

Chapter 11

Creative instrument design

B E Noltingk

Windwhistle, Nutcombe Lane, Dorking, Surrey RH4 3DZ, UK

Abstract. There is a need for creative instrument design. This is a process that should continue through all the stages from the concept of a new way of making measurements to the emergence of a piece of operational equipment. The most significant phase is an early one – the recognition of the possibilities inherent in a new idea – and ways of encouraging such creativity are considered. Conclusions can be illustrated from the history of instrument development, some of it documented in the literature, some from the author's personal experience. A few practical points of advice for innovators include emphasising the importance of the signal to noise ratio in any measurement system.

1. Introduction

The three words of my title are all ambiguous. It may be helpful to give some definition of how I shall use them.

Most human thinking follows a predictable pattern. Each thought can be seen as a natural sequel to what has gone before. Indeed, an orderly existence would be difficult if this were not so. But sometimes ideas appear for which it is difficult to see how they have arisen; there is a gap in the chain of development. Of such ideas it is fair to use the word *creative*, stemming from *creation* that may be defined as making from nothing. It could certainly be argued that the apparent gap is really just evidence of limitations to our understanding of the human thought processes, but creativity is a recognisable concept and I want to use a stronger word than 'original' or 'innovative'. Of course, there are degrees of innovation or creativity and it would be foolish to exclude less extreme examples.

As an *instrument* I include any device that is used for making measurements. Techniques, if they can be thought of independently from the hardware embodying them, are omitted

as not being amenable to design.

Sometimes preproduction activities by a manufacturer are categorised into research, development and design, coming in that order, with the last-named only putting the finishing touches to ideas that have been largely moulded at the R and D stages. Here, however, I am using the word *design* to cover all the stages, even giving a greater emphasis to the earlier concepts.

The creative phase of instrument design discussed here is a small part, in terms of effort, of the total process of producing effective instruments. It is nevertheless of great importance. We naturally think more of situations where the basic concept is original, but it should be remembered that the best design team is alert to innovation in many small ways as well as in a few big ones.

Instruments are used in science and in technology, both to understand nature and to harness it for human requirements. In both applications there is need for a creative approach. There is a cultural need for the continual extension of ways of making scientific measurements. Equally, there is an economic need for new instruments. This can be thought of as part of a global struggle to make the most efficient use of earth's resources. It can be thought of in narrower terms, that the survival of an industrial community depends on its ability to innovate and excel in some competitive fields, as was brought out in the Finniston Report (Finniston 1980). For Britain, competitive fields include instrumentation. It would seem that there is a greater ability in this country to propose radically new ideas than at the production engineering phase of embodying them in commercial hardware; and we should do what we are good at. We need creative instrument design.

2. People and places

2.1. Education for innovation

The debate will no doubt continue as to the make-up of creative people for scientific and engineering work. Instrumentation is like other subjects in the characteristics called for in the people needed to infuse it with new ideas. I have written previously (Noltingk 1969) about their educational requirements. There is a tension between the need to impart a wide range of factual knowledge and the need to keep the learner's mind open and his confidence high so that he is not inhibited from proposing his own radical contribution. The former calls for concise, dogmatic teaching and a long course. The latter object is better served by teaching pupils to question what they are told and by giving them early scope for self-expression. Perhaps the solution is to give the high fliers a different education from that of those destined for supporting roles, but there is the great problem then of recognising early enough which is which.

2.2. Teams

While creative design tends to be thought of as an exercise for the individual, there are advantages in having a group of people working together. New ideas are often shaped best as two people, with some diversity but with sufficient in common to be able to talk easily, think out a problem together; this may be a slower but ultimately more effective alternative to the Brainstorming we discuss later.

If experts in different instrument disciplines are available within the one team, their skills can be brought to bear on a problem as development proceeds and their advice will be on tap at any stage. Optics, electronics, data handling – to name but three topics – are all important and specialised. The all-rounder will combine theory and experiment but should be supported by people with a stronger preference for one or the other. Again, the more creative individual can be helped by a colleague who looks through more conservative eyes.

Moreover a sizable team allows scope for switching effort from one project to another as possibilities and priorities change.

2.3. Applications versus techniques

One way in which instrumentation research is different from most other subjects is the sharp distinction between an 'application orientation' and a 'technique orientation'. In the former, the approach is to start with a need and think of any technique that could solve it; in the latter, an expert in a technique thinks of all the problems it might be used to solve. For application orientation a wider, more general education is desirable while the narrower specialist in the particular discipline will be needed to follow a technique to its limits. An advantage of having a comparatively large team, closely coordinated in their instrument research work, is that it makes possible some combination of the two orientations.

Utterback (1971) has made a systematic study of how new scientific instruments originated (presumably in the USA). He found twice as many arising from application orientation (called 'needs') as from technique orientation (called 'means') and even a much greater ratio for the more commercially successful instruments.

When we come to consider what kind of establishment may be the birthplace of a new instrument, there are four candidates:

Universities or other places of higher learning, Independent laboratories, Instrument manufacturers and Instrument users.

2.4. University laboratories

Kelly (1980) has argued the appropriateness of universities for sponsoring by Government when instrumentation research is needed, reckoning that that agrees with the universities' function of transmitting, interpreting and extending knowledge. It would be fitting for a university that had been involved in fundamental scientific discovery to be concerned also in innovative applications of it, while the small team that is typical of university research is less of a limitation for instrumentation development than in some other fields. The implication is, however, that universities would concentrate on the less productive technique oriented research. And, as we have pointed out, there are advantages in having a larger team.

2.5. Contract laboratories

Rather nostalgic sentiments would welcome the private inventor being reinstated in the place he seems to have had a century ago. If we think of the independent, contract-research laboratory as his successor, then we might be glad at its picking up some role in instrument development. But it is not often practicable for such a body to carry sole responsibility: the long time span from conception to profitability of an idea makes it difficult for an outside body to move in with its own creation, to complete development and then move out again to start up in another field. Moreover, for commercial success, a link must be established with the customer. A more likely part to be played by an independent laboratory is in partnership with a user or manufacturer, who will probably have had the primary original thought and brought it for further investigation to a place where the appropriate expertise can be found.

Lemcoe's work (1976) on high temperature resistance strain gauges perhaps has a relevance to the problem. His studies identified reasons for the inconsistent behaviour of existing gauges and could have meant that devices with improved characteristics became available, though appreciable further effort would have been needed to ensure consistent and continuous manufacture. The original work was done at Battelle, an independent research laboratory, under contract to government agencies, and for reasons of their own Battelle opted not to pursue a course that might have led to commercialisation of a specific product.

2.6. Manufacturers' laboratories

We are left to consider the laboratories of instrument users and manufacturers as potential homes for creative instrument design. A manufacturer should be continually updating and improving his products but this will normally be in small ways – what Finniston calls incremental innovation and Kelly calls evolution. Some instrument firms were actually started in order to produce something novel, and certainly some introduce revolutionary new models. I suggest that nowadays manufacturers tend to be taken up with rather shorter term problems so that they do not have the effort – nor perhaps the more leisurely atmosphere – for radical innovation. This assessment is supported by von Hippel (1976).

2.7. Users' laboratories

With instrument users, an application orientation is natural. They will be aware just how acute is a particular need. If a laboratory has been set up to serve a group of users, it is probable that the whole exercise will have been big enough so that day-to-day pressures for quick results have been removed. As we discuss later, this is likely to provide a good environment for creative thinking. The wealth of instruments originating at places such as BISRA and the Shirley Institute twenty or thirty years ago confirm this expectation.

To summarise, the most likely places to look for creative instrument design are user laboratories if it is application oriented and universities if it is technique oriented.

3. Stages of development

I have already referred to the Finniston Report (Finniston 1980), which dealt primarily with how to promote the 'engineering dimension', but as part of the argument stressed the urgent need for innovation. Finniston distinguishes between 'radical' and 'incremental' innovation. In this present paper, I am concerned more with the former, while remembering that the latter must not be ignored. Finniston also stressed the need to pay attention to market opportunities. That point arises at the very beginning of the development of a new instrument. Market opportunities refer to the present and also to the future. A project for instrument development is more attractive if it has at least the possibility of leading to a 'stretched' version of the first model and also if the expertise built up among staff may prove to be a stock-in-trade for other, later developments. By paying attention to this aspect, there can be some counter to the problem that instrument development tends to be piecemeal, fragmenting the work of a group as they respond to calls for a diversity of techniques.

The stages of a typical development (Noltingk 1969) may be listed as follows:
(1) Problem posed from customer requirements. Functional specification prepared formally or informally.
(2) All conceivable alternative approaches listed.
(3) Theoretical assessment, rejecting most approaches.
(4) Detailed work, theoretical and experimental, on most favoured approach.
(5) Engineering and proving of final design.

3.1. Specification

A functional specification may seem a dull matter to think about when glamorous invention is in the air. But it is very important. Explicitly or implicitly it must be there for new ideas to be judged against. The customer or his spokesman of course has to agree to it but the instrument developer has often more experience in formulating such things and may point out requirements, such as speed of response or operation in a hostile environment, that the customer has taken for granted.

The first numerical stage in preparing a specification is to set down orders of magnitude, notably for what accuracy is needed. A very helpful next stage is to have a 'bracket specification'. The upper limit of the bracket is the performance figure that there is just no point in improving on: the customer will have no use for such a refinement. The lower limit is that below which the customer will have no use for the instrument, no matter how cheaply it is produced and no matter what bonus facilities it provides. Of course, such black and white terms are oversimplifications; there may well be interaction between different features, so that if one is very good another can be tolerated rather worse. But it is a healthy discipline to aim to be quantitative and simple. It will help to sound a note of realism. Most new instruments will be in competition with older ones and there is no point in their being different for the sake of it; any novelty that is introduced must make them better in at least one respect.

Armed with a functional specification, perhaps formally agreed, perhaps just stored in the would-be inventor's memory, the creative phase is reached of proposing ways to meet it. This is the kernel of creative instrument design. We return to analyse it in the next section.

3.2. Early assessment of possibilities

The first appraisal of possibilities is a very critical exercise. On the one hand it is a waste of time to explore in depth – experimentally or even theoretically – ideas that can confidently be predicted to prove unsuccessful. On the other hand there is a long list of technological proposals that have had scorn poured on them by those in high places. No doubt instrument innovations have not been exempt from this. I wonder, when bonded wire strain gauges were first proposed, did some expert say that it was unthinkable that domestic glue and paper should be able to transmit distortions with a consistency of one part in a million?

When the CERL Acoustic Ranger (Morgan and Crosse 1978) was first proposed – a tool that has proved highly successful in locating blockages and leaks in tubes – I quite expected that its signals would be completely swamped by the extraneous noises inseparable from a boiler in a power station. Luckily I made no attempt to veto the exploratory experiments.

Reminiscing further back, we had a fascinating time shortly after it had been established (Greenwood and Williamson 1958) that a conductor, made from Armco iron and shaped to have a very sharp constriction, could exhibit negative resistance. Clearly this had potential as an amplifier to rival the valve and the then newly emerging transistor. For a week or two we worked feverishly to make such devices with an adequate robustness and a usable impedance. After a few man weeks of effort had given little real prospect of improvement, I was convinced we were on a loser and pulled out, but I believe that expert consultants insisted on witholding a firm verdict and it was many thousands of miles of salesmanship later before the project was abandoned.

Such illustrations show how critical it is, for economical deployment of resources, that sound and early judgments should be made. They also demonstrate that the third phase of development (§ 3) is likely to spill over into the fourth. There is much work in the fourth phase and probably more in the fifth, but they are the less creative parts so I pay less attention to them here. Some tension is to be expected at the engineering stage of new instruments. Innovation is rightly encouraged in secondary matters and may be essential to support the primary innovation. On the other hand, a multiplicity of unproved and interacting features coming together in a single instrument could well lead to a prolonged trouble-shooting exercise. A balanced judgment is needed.

There is often some iteration between phases. Difficulties experienced later may suggest re-examination of the functional specification. There is also interchangeability – and this is true of much of science and technology – between the theoretical and the experimental approaches. The electronics engineer often deliberately chooses not to make his calculations precise because he knows it is quicker to change components after making some measurements, than to work out just what he should use in the first instance. I was told that it was the other way round with that wonderful instrument, the mass spectrograph (Aston 1919). Professor (later Sir Charles) Ellis explained in his atomic physics lectures that F W Aston and colleagues were engaged on war work in the First World War. When off duty, they talked of the clever machine to be built when peace came, and so carefully had they thought about every detail that when it was ultimately assembled their mass spectrograph worked immediately, in a manner quite unprecedented for something so new.

3.3. Time taken over development

The length of time over which a complete development is spread may be frightening. Thirteen years ago (Noltingk 1969), I noted a project then active in stages 4 and 5 (§ 3); it is now nearly finished (as it has been for the last seven years)! Utterback (1971) quotes times ranging up to 56 years, but he is talking more of the period between availability of technical information and the development of an instrument based on it. He found that instruments arising from 'needs' had had the information available anything up to 56 years before, with a mean of 12 years. Shorter periods were involved for instruments originating in 'means', a maximum of 10 years and a mean of only 2.

The history of wartime developments (Jones 1975), when complete new projects – some of them would be called instrumentation – were brought to maturity in a matter of months, shows that there is no absolute law forbidding rapid progress. It must depend on patterns of organisation and motivation that are formulated very differently in war and peace.

3.4. Patents

Patenting should be mentioned as one stage of developing new instruments (Kirby 1981). Attitudes to patenting vary greatly. Some larger firms apply for patent protection on every possible occasion; some people reckon that the disclosure that must accompany a patent application gives a greater danger of successful competition than does the absence of legal protection. Disclosure by publication can, of course, give some protection against other people patenting an invention and needs a negligible financial outlay. The common belief is that patent protection should be applied for as early as possible. I suggest that this is often based on a false conceit as to how important the invention is and how hot the competition. If instead a patent application is delayed as long as possible, it will give more time to work out aspects of the invention, it will put off the date when the patent ultimately expires and, most important, it will allow time for a more considered judgment before decisions have to be made about how far expensive foreign protection is desirable.

4. The origin of novelty

4.1. Climate for innovation

In much of science and technology, and indeed in other parts of human society, there is great value in good innovation. Therefore much attention has been paid to analysing how it comes. If new ideas cannot be commanded, at least a favourable climate can be established, in which there is a higher probability of their manifestation. The creative step in instrumentation is no doubt similar to that in other science, and perhaps even more so to that in other technology, where the factor of applicability must not be ignored. Instrumentation is perhaps an extreme example of a field where it can be profitable to draw together scientific matters that are at first sight unrelated.

It is often difficult to analyse the mental processes that create innovation. If we cannot fully understand them, at least we can pay attention to the external situations fostering them. Perhaps it is a case of providing the right stimulus for the subconscious – if we say that all thinking that is difficult to analyse takes place in the subconscious! How does the would-be innovator prepare?

4.2. Prior knowledge

Some assimilation of the existing state of the art is necessary. We have said that too great an emphasis on current techniques may inhibit the imagination needed to introduce radical novelty. However, they cannot be totally ignored. Part of the background must be a broad awareness of what has been tried already; this is a way of appreciating the specification that must be met, and it reduces reinvention. In fact, innovators should not be too frightened of reinventing. They would be in good company: Kelvin found that he had been anticipated in his independent ideas at least 32 times (Lenihan 1979). What is important is to admit the earlier priority if it is pointed out – and to remember that circumstances may have changed so that what was less acceptable earlier may now meet a real need. Information services can now, with the growing application of computer power, be so effective that it is easy to gain a broad overview of a technical situation, at least to the extent that it is reflected in the literature. Innovation for the future should be built on an awareness of the present.

4.3. Brainstorming

'Brainstorming' is often advocated as a source of new ideas. Several imaginative people, preferably with different backgrounds, are posed a problem and invited to toss in any wild ideas that might conceivably be a basis for a solution; there must be a rule that the authors of the most transparent absurdities shall not have them held against them. A great deal depends on the swift judgment of the chairman. It is true that one idea can lead to another, sometimes a bad one to a better one. It is also true that the immense diversity of the techniques that might be applied make instrumentation a promising field for brainstorming. However, I cannot recall any major developments reported as starting this way. Perhaps the approach is really most used negatively as an alibi allowing a report that 'even a brainstorming session including X, Y and Z could only . . .'.

4.4. Suggestions from Nature

Laithwaite (1977) would stress the value of studying Nature as a source book for technological advance. In optics, in electrics, in ultrasonics, detection and measurement abilities of other species have been far ahead of what man developed. I suspect that a recent illustration of modern developments paralleling Nature arises in data handling; it is now thought that the wealth of information received by the eye is processed locally before transmission to the brain, in line with the modern fashion for micro- and distributed computers to supplement mainframe ones. We might expect that when the secrets of animal navigation are finally uncovered they would suggest radically new ways for humans to do it.

4.5. Creativity in the subconscious

If creative proposals emerge from the subconscious, or if they come from the fortuitous conjunction of two ideas that are not at first sight related, then there is not very much deliberate planning that can be done to ensure their occurrence. The aim must just be to ensure a favourable climate. This will go, for one thing, with a comparatively leisured approach to research and design; pressure to get ahead with urgent and obvious developments must not be strong enough to force out the random thoughts that could lead to innovations.

A curious eye must always be kept open to observe new things; it must always be backed up by a memory stored with a list of instruments waiting to be invented. The Courtney–Pratt image dissection camera (Hayes 1956) was conceived when its originator noticed an optical novelty in a Paris shop-window. He had the understanding to analyse its performance and was continuously aware of the need to improve and innovate in his field of very high speed photography, so it was natural for application and technique to be married.

I can speak with some personal experience of flame monitoring. In the early 1960s, it was becoming apparent that the monitors then available were inadequate for multiburner boilers. My own response to the situation was that the only thing that could be done would be a closer study of flames in the general hope that that might throw up something that could be utilised in a novel monitor. In 1967, I attended IMEKO 4, where there was much talk of cross correlation – mainly for flow measurement. The idea must have registered with me that cross correlation did not have to be just an abstract mathematical exercise, because a few weeks later I realised that it could be put to use in a system that would recognise the presence of a flame in a particular region of three-dimensional space (Noltingk et al 1975). The system has been widely used since.

4.6. Three-dimensional flexibility

An isolated point of advice to inventors is that three-dimensionality is a useful thing to aim at. If the behaviour of a device depends on its geometrical shape, there will be added scope for varying its performance and therefore for optimising linearity or some other property. That is true in two dimensions and a fortiori for solid geometry. The Kirby grip hair-pin (though scarcely an instrument) is reputed to have been the most successful invention ever; its action depends on its shape.

4.7. Signal to noise ratio

Another isolated piece of advice to the instrument innovator is to remember that nearly all measurement problems are really matters of signal to noise ratio. At a very early stage of any new proposals, think what noise will be present, noise being defined here as any phenomenon liable to give an output from the measuring instrument that will be difficult to disentangle from what is caused by the true output. You improve your prospects of success by reducing the noise just as much as by increasing the signal.

5. Some examples

5.1. Josephson effect

The potential for a very wide range of new instruments arose with the discovery of the Josephson effect (Josephson 1964). To do justice to the subject would unbalance this article: more than a thousand papers have been published (Giffard et al 1976). It may also be said that, just because the applications have been so wide and so exceptional, lessons from the history of Josephson junctions must be applied sparingly to the creative design of other instruments.

Josephson junctions have been used more in scientific research than in technology. They have served in magnetometers and in the search for gravity waves – vibration amplitudes approaching 10^{-19} m being detectable. Electrical standardisation has also benefitted. Applications in biomagnetic and geomagnetic studies are also reported. Their great virtue is the remarkably low noise that can be associated with them.

They have been used in instruments developed for single purposes rather than commercially available items, perhaps because of problems of reproducibility of the devices themselves.

Josephson junctions belong firmly with the category of technique-oriented instrument development. The vital creative step was the scientific discovery. Application of it to particular situations, while important and demanding, is relatively straightforward.

5.2. Mössbauer spectroscopy
Mössbauer spectroscopy is similar to Josephson applications in being a measurement application of a major scientific discovery. It can be thought of as a technique rather than as giving rise to particular instruments and as such would fall strictly outside the scope of this paper.

5.3. Vortex flowmeters
Vortex flowmeters are an interesting innovation. They appeared suddenly in the British literature; there was no mention of them in a 1974 Colloquium on 'Advances in Flow Measurement' while the 1977 Symposium (Institute of Measurement and Control) 'The Application of Flow Measuring Techniques' refers to them as well established (Lomas 1977). Looking abroad, however, to USA and Japan, we find that there was a more gradual development. (Yamasaki and Kurita 1967, Yamasaki and Rubin 1971 and White et al 1971).

Leonardo da Vinci observed vortex shedding in 1513 and Strouhal a century ago established the proportionality between fluid velocity and frequency of shedding (Lomas 1977). In 1954, Roshko (1954) pointed out the possibility of measuring flow this way while in 1960 Shiba (1960) made the first practical attempt at it. The creation of this design of instrument was thus spread over a decade and shared between several people. There seem to have been three identifiable creative steps: 1. The realisation that this long-known phenomenon could prove advantageous in a field where several other techniques were already available. 2. Choice of a particular shape of bluff body so that larger vortices were produced. 3. Improvement in the transducers used to detect the vortices that had been shed, and so to initiate the pulses whose frequency was the signal indicating flow velocity.

5.4. Capacitive strain transducers
I can write out of the experience of participating in the development of the CERL/Planer high temperature strain transducer (Noltingk et al 1972). Its history can usefully be considered in terms of the development phases outlined previously (§ 3).

There was a need, particularly in the CEGB , to measure slowly varying strains on plant such as turbine casings, at temperatures up to 600°C. The essence of the functional specification, therefore, was simply to improve on existing resistance strain gauges in matters such as slow, unpredictable drift – which can be described as reducing this particular component of noise. Phases 2 and 3 took place over an extended period and led to the decision to see whether a capacitive transducer could out-perform a resistive transducer.

It was an important step forward to recognise that accurate measurements could be made of capacitances less than 1 picofarad. Blumlein had done this twenty years before but his achievements do not seem to have become universally known and part of the specification in parallel work (Gillette and Mullineaux 1969) was that the transducer should include more than 10 pF. The electrical measurement is much easier if neither electrode of the capacitor is earthy, but given that concession it is reasonable that the circuit designer should shoulder part of the problem in order to make things easier for the transducer designer.

While favouring a capacitive principle, I personally did not like an approach using a variable air gap, and thought in terms of some compressible solid dielectric. This was because memories of earlier transducers (Carter et al 1945) had led to a general bias instead of a rational analysis of their weaknesses. Fortunately, McLachlan was not inhibited by this attitude and used his scope for design in three dimensions to come up with the configuration that is at the heart of the device. That was development phase 4. Phase 5 included worries about materials and bonding methods, needing a continued creative approach.

When a version suitable for 300°C was described, I recall one of my colleagues remarking that it would be a very long step before the 300°C had been stretched to 600°C. But the very next report from the developer showed how plastic could be replaced by ceramic. Sometimes creativity works quicker than we expect.

6. Conclusions
This is a rather unstructured and anecdotal paper. It is difficult to draw firm conclusions. I believe that a closer examination of the situation and a brief study of some individual cases have confirmed the great importance of a creative approach to instrument design, an approach that must be continued beyond the initial concept until a piece of equipment is fully operational.

The right human environment is more difficult to specify – corresponding perhaps to the large variance in the relevant properties of humans. It is exceptional for a major instrument development to be attributable to only one person.

References
Aston F W 1919 Positive Ray Spectrograph
Phil. Mag. **38** 707–14

Carter B C, Shannon J F and Forshaw J R 1945 Measurement of displacement and strain by capacity methods
Proc. I. Mech. E. **152** 215–21

Finniston Sir Montague 1980 (Chairman) *Engineering our future* Cmnd 7794 (London: HMSO)

Giffard R P, Gallop J C and Pethey B W 1976 Applications of the Josephson effect
Prog. Quantum Electron. **4** 301–402

Gillette O L and Mullineaux J L 1969 Development of high temperature capacitance strain gauges
ISA Trans. **8** 52–61

Greenwood J A and Williamson J B P 1958 Electrical conduction in solids II
Proc. R. Soc. A **246** 13–31

Hayes R A 1956 High speed image dissection photography
Industrial Image (Ilford) **1** 3

von Hippel E 1976 The dominant role of users in the scientific instrument innovation process
Research Policy **5** 212–39

Jones R V 1975 Epilogue (to a discussion on the effects of the two world wars)
Proc. R. Soc. A **342** 549–54

Josephson B D 1964 Coupled superconductors
Rev. Mod. Phys. **36** 247–51

Kelly A 1980 *'Government research at universities', R and D Soc. Symp., London April 1980* (London: Research and Development Society) pp 14–23

Kirby P L 1981 (Chairman) Scientific patents, Inst. Phys. Conf.
Phys. Bull. **32** 100

Laithwaite E R 1977 Biological analogues in engineering practice
Interdisciplinary Science Reviews **2** 100–8

Lemcoe M M 1976 Development of electric resistance strain gage system for use to 2000 F
Advances in Instrumentation **30** 6–9

Lenihan J 1979 *Science in Action* (Bristol: The Institute of Physics) pp 91

Lomas D 1977 Vortex, Turbine, Orifice – which one do I choose?
Flow-Con 77 Inst. M. C. Symp. on the application of flow measuring techniques (London: Institute of Measurement and Control) pp 15–44

Morgan E S and Crosse P A E 1978 The acoustic ranger, a new instrument for tube and pipe inspection
N.D.T. Int. **11** 179–83

Noltingk B E 1969 Education for Innovation in Measurement Techniques
IEE Conf. on Measurement Education Conf. Pub. No. 56 pp 45–9

Noltingk B E, McLachlan D F A, Owen C K V and O'Neill P C 1972 High stability capacitance strain gauge for use at extreme temperatures
Proc. IEE **119** 897–903

Noltingk B E, Robinson N E and Gaydon B G 1975 A new approach to flame-detection in multi-burner boilers
J. Inst. Fuel **48** 127–31

Roshko A 1954 *NACA Rept. No. 1191*

Shiba H 1960 *Trans. Jap. Soc. Shipbuild.* **97**

Utterback J M 1971 The process of innovation
IEEE Trans. **EM-18** 4 124–31

White D F, Rodely A E and McMurtrie C L 1971 The vortex shedding flowmeter
ISA Conf. on Flow (North Carolina: Instrument Society of America) pp 967–74

Yamasaki H and Kurita Y 1967
6th Proc. Jap. Soc. Inst. Contr. Eng. Conf. p 389

Yamasaki H and Rubin M 1971 The vortex flowmeter
ISA Conf. on Flow (North Carolina: Instrument Society of America) pp 975–83

Chapter 12

Future instrumentation: its effect on production and research

R Shaw

Engineering Department, Imperial Chemical Industries PLC, Petrochemicals and Plastics Division, Welwyn Garden City, Hertfordshire, UK

Abstract. An introductory section gives a general profile of the instrumentation and control industry in the UK, making particular mention of the increasing use of microelectronics-based systems. Subsequent sections cover in greater detail sensors, analysers, signal transmission equipment, and instrument hardware; the growing tendency for equipment to be comprised of basic hardware and adaptable software is noted.

1. Introduction

The last 100 years has seen instrumentation move from the simple principles of the dial thermometer and electrical measuring instruments of the 1880s, to the electronic digital control of the 1980s. Throughout this period the most constant aspect of the technology has been its state of rapid change (see Appendix 1).

To the instrumentation manufacturer the evolving technological environment has represented an opportunity for profit and growth. But for the user, in production or research, the major benefit of instrumentation is as a financial amplifier. Within the national capital investment in plant and machinery on average some 5% is spent on instrumentation and control. In exchange for this the user expects product quality, higher plant utilisation, material and energy savings, better manpower productivity, and speedier accurate data reduction. There is little reason to doubt the claims that for systems which are well

specified and designed the pay-back can be achieved in under one year's operation.

Research Departments have long been the home of sophisticated instruments because of their need to measure esoteric variables accurately. Rising staff costs have brought an increasing awareness of the need for electronic data capture and automatic sequence control of experiments. The steady penetration of data links and information networks is beginning to indicate the future organic and neural nature of instrument and information systems in the laboratory.

2. Profile of the industry

2.1. The market

To provide the users with their systems the supplier of instrumentation and control equipment competes in an aggressive and fast-moving international market. As a UK industrial sector instrumentation is relatively healthy: of the UK market for instrumentation and control, which in total exceeds £500 million, half is supplied by imported equipment but the industry nevertheless achieves a modest positive trade balance.

The evolving commercial environment can be seen at work in figure 1 which is a pie chart of sales in key market sectors together with the annual rate of growth. Oil and petrochemicals show the expected declining growth rate while new energy sources and manpower-intensive industries are the currently favoured areas.

2.2. Ownership

A significant feature of the UK instrumentation industry is the dominance of large overseas-owned companies; the bulk of UK ownership is concentrated among the medium and small firms. Another strong influence is the weighty purchasing power of the major plant and process contractors and large nationalised

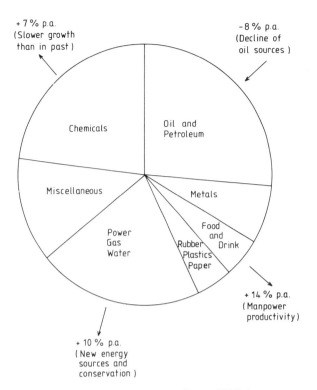

Figure 1. UK users. Total UK market ~ £500M.

industries. The nett effect of overseas ownership and powerful but traditional customers can be to encourage conservatism at the expense of innovation and equipment developed abroad is often imposed on the UK market in preference to bright ideas conceived here.

2.3. Capabilities

Despite the harshly competitive climate (or perhaps because of it) the available variety of sensors, instruments and control systems is truly impressive. From abrasion and adsorption, through dustiness and eccentricity, past hardness and hygrometry, by way of moisture and noise, via thermography and thickness to weight and width, the product lists of the instrument industry catalogue a bewildering array of equipment to measure, control and test. Each month the list grows as new techniques probe the hitherto unmeasurable and regulate the previously uncontrollable. While Appendix 1 is a fascinating backward glance, figure 2 shows what interests the users now.

2.4. Microelectronics

The strong growth trends in analysis techniques and digital

systems are evident in figure 2 and both of these have been strongly stimulated by recent solid state technology and microelectronics developments. The pattern of steady penetration by microelectronics is equally evident in scientific instruments where it provides significant breadth of capability, much enhanced diagnostic facilities, and flexible communication interfaces unimaginable at the price 10 years ago. Two features in particular relating to the use of microprocessors in instrument equipment are worth comment. Firstly it is unlikely that the total power or versatility of the microchip will have been fully utilised in any unit design. Despite this, the economics of microelectronic technology will have been attractive to the supplier, especially bearing in mind the easy way in which later enhancements and design updates can be incorporated with only a small development overhead.

Secondly, so great has been the speed of advance in hardware microcircuit technology that related techniques like sensors and software now represent constraints on progress. What then is the sensor scene?

3. Sensors

3.1. Present status

Understanding rests on sound information whose foundation, in turn, is accurate measurement. In a study of its instrument and control technology a major chemical company found that over

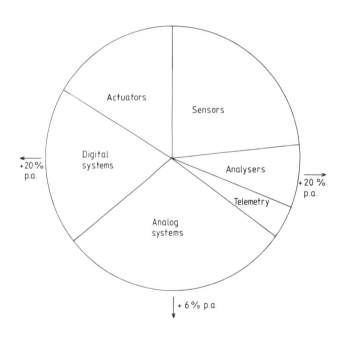

Figure 2. Equipment purchases (by value).

133

50% of its total instrument maintenance effort was devoted to process measurements. From a range of typical plants a composite sensor population profile was developed (figure 3). This profile only serves to emphasise how dominant still is the traditional type of sensor (and incidentally the traditional instrumentation knowledge and skill required to service it) and the relatively insignificant impact as yet of microelectronics on sensors. Sadly, it also indicates how very few are the examples of non-invasive measurements (categories marked with asterisks).

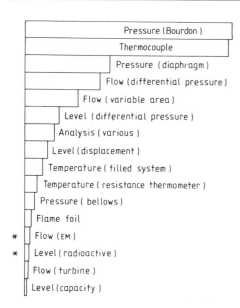

Figure 3. Sensors (relative populations). Asterisks denote non-invasive measurements.

Discounting for the moment analysis equipment and considering total maintenance effort for each type of sensor, the lion's share of support goes into differential pressure, level and flow equipment followed by thermocouples and resistance thermometers, variable-area flowmeters and pressure gauges, in that order.

3.2. User needs

But this is not quite the problem priority order as seen by users. They want a faster-responding and more accurate temperature device than the thermocouple, ideally using a non-contacting technique to avoid the numerous physical and heat transfer problems associated with thermo-sheaths. Next comes flow measurement with a call for non-invasive techniques (incidentally offering energy savings) to cope with a wide range

of flows, temperatures, material density variations, viscosities and compositions. Level measurement sits in third place with special difficulties caused by dust, obscuration, coating and sticking, intrinsic safety, and again the problems of invasive measurement.

The placing of these three measurements at the top of the list does not imply that no problems exist with any others: all variables can present difficulties.

3.3. Credibility

As if these physical troubles were not enough it appears that control-room operators have a low level of belief in their instrument readings. They make continual calls for instrument checks on the plant, less than 20% of which show any fault, but cause significant interruption to production. This finding alone points to the need for more comprehensive built-in proof testing facilities in measurement instrumentation to facilitate instant diagnosis by the operator in the control room or even automatically.

3.4. Reliability

Analytical techniques do exist to determine at the design stage the ways in which an instrument might fail either partially or catastrophically. In the latter case the failure rate is found by summing the observed failure rates of the modules from which the instrument is constructed. Partial or potentially unrevealed maloperation analysis calls for a detailed study of the various modes of partial failure like amplifier drift or component aging, to generate an error probability function. Comparison of the two analyses will suggest the frequency of routine checking, calibration or replacement.

In all cases however, the cost of designing for higher reliability must be balanced against the operational penalty of failure.

3.5. Ergonomics

However reliable the instrument, its value in use will depend on how effectively it communicates and interacts with the operator; its ergonomics.

Research studies and rules of thumb on the topic are legion. Analogue indication is best for qualitative information and digital for precision. The height of digital characters should be 25% larger than that of associated labels and have a ratio of stroke width to height between 1:6 and 1:10 depending on contrast with background. 7 × 5 dot matrix displays are superior to seven-segment characters and the error rate for a green display may be up to three times that for red. Analogue indicator scales should not be multiple or non-linear and numbering should increase clockwise. Fixed-scale moving-pointer indicators with coloured markings or bands to indicate danger conditions can speed recognition of abnormal conditions up to four times.

The vast possibilities soon to be offered by time-shared flat-screen electronic displays with auto-zeroing, auto-ranging and selfdiagnosis in addition to multifunction presentation will re-emphasise the golden rule 'question the need for the information and keep it simple'.

3.6. Invasiveness

In relation to the generally invasive nature of measurements (figure 3.), the instrumentation survey revealed that when trouble does occur, causes predominate which are associated with the sensor interacting with the process such as chokes, coatings, corrosion and damage. This suggests that Doppler flow measurement, hoop pipe-stress pressure sensing, ultrasonic process signature identification, and ultimately the new purely optical measurement techniques will be important as the basis for further development in sensors.

Meanwhile, putting several sensors together in a cluster with only one tapping point into the process could offer advantages. Cheaper installation, greater reliability through redundancy, less process energy loss and simpler maintenance would be some of the benefits. Some redundancy of this kind must also be the starting point for measurement validation checks and automatic fault-testing diagnosis procedures.

3.7. Pressures on development

The direction of future sensor development may be influenced by two strong and relatively recent pressures. The first arises from the common ground between the objectives of industrial and research users. The former emphasise high reliability and repeatability with accuracy about 1% or better, while in research applications, which can be more esoteric, the emphasis is on higher accuracy and flexibility of function. The instrument suppliers' response to both requirements is to explore and refine newer techniques, like vortex and magnetic flow metering, while extracting more data from traditional sensors using electronic techniques.

Fortunately techniques like signal correlation, comparison of multiple input signals, digital filtering and smoothing, variable-rate sampling and Kalman estimators offer much higher information content from existing input signals. These new sophistications will put pressure on the instrument makers to link more closely with manufacturers of microcircuits for the specialised low-volume customised chips and masks which they need.

It is from the solid state technology of the chip manufacturers that the other pressure on sensor development arises. The massive investment by vehicle manufacturers into sensor-based automobile electronics completely dwarfs the most optimistic estimates of instrument industry R & D investment. The 1981 demand by General Motors for integrated circuits and solid state sensors will be £188 million with an estimated worldwide 1984 market demand of £600 million. The present sensor specification in automobile and consumer applications may well be modest accuracy, strictly limited flexibility and cheap interchangeability but the huge investment and resulting

steep learning curve should rapidly put this technology in a position to exert a powerful influence on industrial instrumentation.

3.8. Optical sensors

Many instrument applications are in hazardous and potentially explosive environments. Intrinsic safety and barrier techniques are well established but not cheap. Possibilities exist for pulsing the power to sensors below the explosive limit or deriving power locally from available plant energy. But in the longer term, a totally different technology based on optics may obviate the safety problem altogether and have other more general attractions.

Numerous ideas for the manipulation of optical information have lain dormant since the mid-nineteenth century but the convergence of fibre optic transmission, new solid state materials, and the need for yet faster data-processing circuits has led to a spontaneous interest recently in optical sensing among workers in universities, manufacturers' laboratories and one or two major instrument users.

3.9. Optical techniques

Optical sensing is based on the principle that the physical variable to be measured causes a change in the optical properties of the sensor and this change in turn affects some property of the light passing through it. Typically this could be phase, amplitude, polarisation state, birefractive indices, or interference fringe effects. An example is temperature-sensitive semiconductor phosphors with spectral cut-off proportional to temperature.

Pressure sensing has been achieved by phase change detection relative to a reference wave down a long length of single-mode fibre or by using optics to detect sensitively the movements of conventional diaphragms. Optical detection of vortex shedding from an immersed bluff body can be used to measure flow, while multireflection splitting and interference pattern detection may ultimately be a more general principle which can be applied to the measurement of many variables.

Attention is also being devoted to the development of multiport data rings and networks. Thin transparent films on a transparent substrate of lower refractive index can be prepared as low-loss film and stripe light waveguides. These act as filters, modulators, switches, power splitters and light beam deflectors. Combinations can form complex circuits for signal processing and switching, from which optical sensors with integral computing capability may finally develop.

The Department of Industry is now taking an initiative in the formation of a UK Association to further information exchange on the whole of this developing area.

Sensors, therefore, still seem to be the Achilles heel of the technology. The users want more reliable sensors even at the expense of accuracy, and tend to focus on total life cost, while contractors are more concerned with initial capital cost. Developments described above, together with emerging computer-aided design methods, may offer the hope that sensors will soon take on an overdue new look.

4. Analysers

4.1. Evolution

Analysis in the last two decades has moved steadily from the laboratory to become a production measuring technique. Early applications tended to be little more than research laboratory techniques applied to samples of production material from some distant plant site. Usually the displacement of the sample and the measurement in both place and time made closed-loop control of quality impossible at that stage.

Gradually the introduction of electronics began to automate the laboratory techniques making them faster, more accurate and reproducible. As a result quality, which for so long has been an inferential measurement, can increasingly be directly measured. By installing specially protected cubicles on the plant, a microcosm of the laboratory environment is now provided close to the sample point, bringing a step nearer the ideal of real-time closed-loop quality control. The latest solid state sensor materials and microelectronics make this close to fulfilment.

4.2. The market

Production-based analysis equipment is almost doubling in quantity every five years and even at the present time absorbs some 20% of instrument support effort. The UK market for analysers is a healthy £25–30 million per annum with a buoyancy derived from the increasing cost of manual analysis and the benefits which cost-conscious users get from better product consistency. Safety and reduction of environmental pollution are further strong pressures.

4.3. Support problems

Analysis instrumentation, however, does pose special difficulties. For the supplier it is, in general, a low-volume high-technology activity and to the user it represents complexity with high installation and support costs. With older equipment roughly half the support work is relatively inconsequential, involving routine checking, cleaning and chemical refilling, all of which leave the equipment in virtually the same state as before.

4.4. New materials

Newer designs offer many more built-in self check and selfdiagnostic features, and in addition have simpler and more reliable measuring elements based on solid state materials which are highly sensitive to specific chemicals, for example, the use of lithium–tantalum solid state detectors to measure vinyl chloride to 0.2 PPM. Also, many of these small-area semiconductor composites readily interface to microelectronics. Progressive surface poisoning can reduce useful life; however this is being improved using theoretical studies of surface chemistry.

4.5. Laser techniques

Both in the laboratory and on plant the increasing cheapness of lasers is steadily widening their application.

Raman spectroscopy methods like CARS (Coherent Antistokes Raman Spectroscopy) and SIRS (Stimulated Inverse Raman Spectroscopy) are recent techniques. One or other of these radiations is emitted when the frequency difference of two coherent beams equals the Raman frequency of the trace molecules of interest.

Another example, the laser environmental atmosphere monitor, joins together spectroscopy (measurement of environmental back-scatter at different frequencies) and electronics (microprocessor control of laser scanning and signal conditioning) and radar-imaging (computer-driven image enhancement and pattern analysis).

4.6. Microcomputer impact

Because of the ubiquitous microcomputer, standard laboratory and scientific analysis equipment now has enhanced speed, accuracy, reproducibility and flexibility. Features like easily selected automated sequence options, signal conditioning, automatic calibration and diagnostic routines, and a range of data-communication facilities to other laboratory equipment are standard. Complex microchips can perform Fourier transforms virtually in real time and together with very-low-noise amplifier techniques imported from radio-astronomy they are now making available infrared multigas analysers with Fourier transform facilities and fast wavenumber multiplexing for real-time absorption analysis of several constituents.

Infrared analysis, which was purely qualitative until the 1960s now offers real-time quantitative measurement covering more than 300 gases. Furthermore, attenuated total reflectance techniques have recently extended the measurement capability to rubbers, plastics and semi-solids using robust calcium fluoride or zinc selenide cells with path lengths less than 0.1 mm.

Speed improvements are particularly impressive: UV and visible spectrum analysers can now span wide-range wavebands in under one second and medical multitest analyses which would have taken two days are printed out in as many minutes. At the other end of the size scale, huge computer-based engine test facilities are similarly reducing testing times in the car industry.

4.7. Sampling

A significant proportion of manpower used in analysis is either in the extraction and preparation of samples or the maintenance of sampling systems, and as analysis techniques become slicker, the pressure to improve productivity in the sampling area will grow. There are welcome trends towards in-line non-invasive analysis techniques and these will increasingly involve fibre optics and optical sensors on the lines described earlier.

The newer solid state materials with high sensitivity may simplify the task of sample preparation but clustering sample points in the plant to provide a group of analysis work stations in a protected environment can give significant productivity improvement. There are now prototype examples of sampling by the use of simple microcomputer-controlled robots in just such a plant-based group of analysers.

4.8. Product quality pressures

Finally a further pressure towards better product quality control is evident. The increasing sophistication of microcomputer-controlled processing equipment at the factories of the customer is revealing batch-to-batch quality variations hitherto averaged out by the relative crudity of the processing machinery.

So analysis is in some respects still the Cinderella of instrumentation, although relevant and interesting techniques abound. Perhaps the awakening influence of investment money will now allow the necessary technology transfer to take place.

5. Signal transmission

5.1. The standards jungle

A healthy current buzz-word is 'digital networks'. Following a surprisingly stable period of analogue instrument transmission standards based on 3–15 psi, 0–10 mA and 4–20 mA, the present digital scene is a jungle. The growing demand for more sensing, monitoring and distributed control systems has spawned equipment-based multicomputer buses, and lately PROWAY, the IEC-sponsored international standard for process control.

Small wonder that although the latter is being studied by instrument manufacturers it is without much enthusiasm. Perhaps they are watching the emerging consensus on protocol standards by the International Telegraph and Telephone Consultative Committee (CCITT) or the efforts by major systems suppliers like DEC, Intel, Xerox, and Hewlett Packard to introduce a *defacto* standard for smaller-scale local area networks (LAN).

5.2. Open systems interconnection (OSI)

The CCITT recommendations are based on the concept of open systems interconnection, in which networks of dissimilar computers, terminals or peripherals of different manufacture will be interconnected and work together without technical incompatibilities. A phased introduction of this environment is envisaged, initially at the physical layer then through the data, network, transport, session and higher application layers of protocol harmonisation.

The wide availability of cheap protocol conversion microchips, like X25 for the network layer, are expected to speed the process of standardisation and force digital control equipment suppliers towards compatability.

5.3. Local area networks (LAN)

LAN arose as the huge office automation market was recognised and it was discovered that the bulk of information is used within 1 km of its source, There are nearly as many networks systems as suppliers but Ethernet may now be gaining a lead.

Adopted by Xerox, DEC, Intel, Hewlett Packard and Siemens, Ethernet allows the interconnection of over 1000 pieces of equipment on a local site area by a coaxial cable carrying information at 10 Megabits/s. Fibre optic cabling is also envisaged.

Alert research groups will not miss the significance of these local networks within a research complex for the manipulation of corporate research data together with data gathering and control at remote experimental facilities.

5.4. Industrial data links

Meanwhile on a multimillion pound plant which may have 40 000 instrument signal connections, multicore cabling can cost hundreds of thousands of pounds and make multiplexed systems economically attractive. Hazardous explosive environments add further complications and require equipment to be flame proof, or intrinsically safe or protected by individual barrier units on each measurement line. All these are expensive. Recent intrinsically safe plant-based systems therefore look interesting especially when they use cheap twisted twin cable and serial data transmission.

5.5. Fibre optic links

Coaxial cable offers higher data rates like 24 Mbits/2 km but even this must concede advantage to fibre optic data links with 100 Mbits/5 km in the visible light range and losses of less than 3 dB km^{-1}. Optical fibres also look good both for process and laboratory or office environments. They are corrosion free, tough, heat resistant, inherently safe and immune from electrical noise, electrically insulated and have enormous information capacity.

Two types of multimode fibre are used industrially. Step index has a discreet step in refractive index at the core–cladding interface and is usable up to a few kilometres before frequency dispersion caused loss of bandwidth. Graded index, with graded refractive index from core to cladding, extends the bandwidth to above 500 MHz km^{-1} for a 70 μm doped silica core material.

As fibre optic cable use grows, the possibilities offered by multicolour (frequency domain) transmission of multiplexed signals as well as high-impedance couplers to ring and network configurations, suggest that in the future data transmission will be optical, perhaps from passive optical sensors.

6. Instrument hardware

6.1. Trends

In process instrumentation the well established trend from pneumatic equipment to electronic equipment (see figure 4) has now been overtaken by the migration from analogue to digital systems so that to talk about instrument equipment is now to talk about hardware and software. Recent price comparisons for a typical 40-loop control installation showed pneumatic equipment costing £38 000, electronic analogue at £47 000 and electronic digital at £35 000.

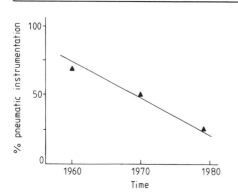

Figure 4. Trend towards electronics.

Not surprisingly the use of microcomputers in the instrument industry is growing at over 35% per year. To assure their supplies and influence the design of microchips, instrument manufacturers are working with or purchasing the chip makers: Honeywell with General Instrument, Foxboro with Integrated Circuit Transducers, and Taylor with Digimetric.

6.2. Micros everywhere

Although initial development costs can be unexpectedly high, microprocessers offer significant benefits to the manufacturer. What was complex hardware design is now a software programming task performed on a sophisticated development system with high-level languages and error correction so that design changes can be made by reprogramming. At the hardware level because there are many fewer components and interconnections, power supplies and packaging costs are reduced and reliability improved. Finally, quite apart from add-on product features which can be accommodated at a relatively low extra cost, later product evolution can be accomplished by program alteration only.

Microprocessors are embedded in transmitters, controllers and recorders and allow many previously 'up-market' features to be offered as standard. Compensated scale adjustment, auto-calibration and -zeroing, power level compensation, range of outputs, multitest diagnostics, high-resolution displays, multimode algorithms, plug-in capability enhancement modules and digital communication interfaces are just some of the features which characterise the new generation of instrumentation. An interesting trend is the quotation of mean time between failure (MTBF) and guaranteed life performance by some manufacturers. This could become a competitive selling point.

6.3. High integrity

Steadily cheapening hardware and relatively easy electronic interconnection has led to the use of multicomponent or multi-unit systems when high integrity is important.

Apart from the normal precautions like minimum component count, proven hardware reliability history, conservative rating operation, initial burn-in and exhaustive system tests, it is possible to improve expected reliability by employing redundancy. Parallel redundancy will improve individual reliabilities from 0.9 to 0.99 for the parallel system. Standby parallel redundancy with switched-in operation of the standby improves reliability from 0.9 to 0.9948. Majority voting (for example 2 out of 3) will give enhancement from 0.9 to 0.972.

Ideally, redundant systems should be designed to avoid faults which might be common to all channels and for example should use different means of signal sensing and control output processing to this end.

7. Conclusions

For long the Cinderella of instrumentation, sensors are finally succumbing to new technological advances in sensor materials, optics and microelectronics. The effect of microcomputers in sensors and instrumentation has been to expand the range of functions available from a single device and to enable modules readily to be linked together in communicating networks.

Always responsive to commercial pressures the instrument industry now seems to have the technological tools to approach the ideal specification of accuracy, reliability, flexibility and non-invasiveness – until the goal posts change again!

Appendix 1

50 years of non-stop development

1928	Reset in electrical controllers
1929	Negative feedback in pneumatic amplifier
1930	Flow ratio controllers Photoelectric cell and control use
1931	Electron microscope UV radiation detectors Flame analysers
1932	Thermistor Precision RID
1933	Variac power transformer
1934	Derivative action used
1935	Cascade control
1936	Piezoelectric elements for vibration
1937	Program controller for batch drying
1938	Strain gauge
1939	Moisture content of gases measured
1940	Three-term (proportional, integral, derivative) controllers

1941	Load cell weighing	1962	Rotary pneumatic motor
			Fluidic amplifiers
1942	Stepless proportioning electric control		Voice recognition
1943	Electronic strip chart recorders	1963	Executive control programs
1944	Valve sizing coefficient established	1964	Digital flowmeter
1945	IR gas analyser	1965	Bang–bang motor control
1946	First digital calculating computer	1966	Thermal mass-flowmeter
1948	First transistor	1967	Minicomputer control
1949	Random access memory used	1968	Lasers
1952	Gas chromatograph	1969	Fluidic controller
1955	Ultrasonic flowmeter	1970	Programmable logic controller (PLC)
1956	Capacitance pressure sensor	1971	LED displays
1957	Ultrasonic level sensor	1972	Visual display units (VDU)
	Dead-time control used	1973	Solid state pressure transducers
1958	Analog–digital converters		Opto-isolators
	First process computer control	1974	Microprocessors
1959	Magnetic flowmeters	1975	Distributed control concept
	Intrinsic safety concept	1976	Colour VDUS
1960	Stepping motor drives	1977	Microprocessor pressure sensor
	Fluidics	1978	Fibre-optic data transmission
1961	Direct digital control		Very-large-scale integrated circuit
	Fibre optics		
	Integrated circuits		

Chapter 13

The literature of instrument science and technology

P H Sydenham

Professor and Head, School of Electronic Engineering, South
Australian Institute of Technology, Pooraka, South Australia

Abstract. Characteristics of instrument science are outlined to
provide an introductory background for appreciation of the
literature of the subject. Various forms of literature are outlined,
including discussion of the diverse terminology used. An
example is used to illustrate the problems of retrieval that can
arise because of the multiplicity of applicable classifications.
Location of book, and of journal, material is reviewed.
Appendices provide examples of standard lists of terms, terms
used in searching, library catalogue numbers, relevant reference
sources and related data bases.

1. Introduction

Selection, design, adaption and modification of instrumentation
involves numerous interrelated decisions. Sound decisions are
based on the possession of sound knowledge.

Each new encounter requires the instrument technologist to
create, and be part of, a maturing design process by which
factual information is assembled, sifted and ordered.

Many kinds of questions can arise – does a suitable sensor
already exist; where can it be purchased; are the specifications
of an instrument realistic; has research been reported on the
principle; what performance figures can be expected or how can
the instrument be assembled more inexpensively?

There exists a vast quantity of information on instrument
science and technology of which much is expressed in the
printed literature. The problem is how to locate it in an efficient
manner.

This contribution to the Instrument Science and Technology
series serves to provide an introduction to the relevant literature
to assist suitable placement and retrieval of published accounts
of principles and practice of instrument systems. Greater depth
of explanation, and of lists provided, is available in volume 2 of
Sydenham (1982).

2. Nature of the literature

2.1. Features of instrumentation

Appreciation of the literature of instrumentation is gained by
considering the dominant features of the practice of instrument
science and technology.

Instrumentation is widely used: it is to be found applied in
almost every field of endeavour.

Innumerable variations on the wide range of fundamentals
arise for the vast spectrum of applications.

Capabilities, relevant experience and training of those
reporting and applying instrumentation vary widely.

In a capital value sense, instrumentation is generally only a
minor part of the whole system in which it plays its vital role.

Also highly relevant is the lack of a strongly defined
teaching/training discipline that would generate a generally
accepted and widely known structural classification for the
fundamentals of instrument science and technology. In
consequence there are no obvious knowledge sub-sets in which
to report or seek contributions.

2.2. Publication outlets

Information is published in many forms. Each has its own kind
of value.

Original contributions of fundamental importance to theory
or practice are usually, but not exclusively, to be found in the
primary journals, Melton (1978). This material is usually subject
to expert review by referees. In the science based disciplines

there are many tens of thousands of *titles* which are published at frequencies ranging from weekly to annually. New material is produced at a rate vastly greater than direct personal searching could follow.

In principle, but not exclusively so in practice, *secondary* sources republish primary information. Their value is that they express knowledge in a more applicable and collected form. Quality of material is most variable.

Much of primary and secondary information is published by impartial bodies such as professional organisations. There also are numerous other publications produced by trade houses, governmental institutions and academic institutions, examples being *in-house* trade journals, product application notes, instrument manuals, internal reports and technical newspapers.

Additionally useful sources are conference proceedings, student theses, patent documents and standards of the specification kind. Each of these sources has its own publishing and accessing network.

Technical books play an important role in that they generally present collected knowledge on specific topics in a structured manner which is tied into the primary knowledge sources. They often contain extensive bibliographies on their subjects. The knowledge presented is, by virtue of the time needed to collect, order and publish the information, at least one or two years behind first publication.

2.3. Publication format

The bulk of information is published in the printed paper medium. There is a trend, however, for much of it to now be duplicated in microfilm and microfiche form. A growing amount is now not appearing in the printed form, being available only in computer compatible or micro form. At present the well known computer based retrieval systems contain only abstracting information, not full text.

A significant proportion of library holdings are formed of reference material in which can be found numerous guides to the literature and means to establish where contributions are to be located. The assistance of a librarian is recommended when seeking information: an account of this brevity cannot hope to replace a trained librarian.

In summary, there exists a bewildering array of potentially useful sources of which only a part is traceable through ordered librarianship. A succinct introduction of relevance is that provided in Melton (1978).

3. Terminology

Information retrieval by *browsing* through original material makes use of many recognition features – words, illustrations, mathematical formulae – but is invariably too slow for searching the literature.

Realistic retrieval procedures are almost exclusively based on the principle of tracing linguistic *terms* used by authors in their titles, abstracts and *index keys*. Consequently the choice of terms written into abstracts, and used in searching, is of paramount importance.

The diversity and uncoordinated multiplicity of the sources of generation (and regeneration!) of instrument concepts has brought with it a highly varied and inconsistent nomenclature. It is only in recent times that standardisation of terms has begun to be accepted; they are not, as yet, widely used nor broadly developed.

Appendix 1 is a selected list of published standards on terms and definitions used in metrology.

The wide use of synonymous terms makes literature retrieval difficult. Apparently obvious terms often lead nowhere – *metrology*, *measurement* and *instrumentation* in particular. The name *instrument science*, for example, might be indexed using at least ten alternatives such as, *measurement systems* and *measurement physics*.

Terms describing the *process* of measurement exist in great abundance (for example, assay, appraise, analyse, inspect, test, quantify).

Measurement, and consequently instrumentation, in specific disciplines often has its own name – anthroprometrics, archaeometry, biometry, faunistics, metallography, psychrometrics, optical tooling. Even the fundamental term *metrology* does not have a singular universal meaning.

Many data bases require the user to list terms that are recognisable by the computer. It is, therefore, necessary to be familiar with the jargon of instrument science to assist efficient selection of material. For example, the thesaurus for the widely used INSPEC data base contains over 8000 terms of which some 650 (which are distributed throughout the whole set) are relevant to instrumentation interests. Appendix 2 is a sample from this source: it is provided here to promote appreciation of the great variation possible for use in a search.

4. Classification

4.1. Library systems

The basis of literature placement in, and retrieval from, a library system is the *numerical classification*. Instrument science material poses unusually great problems due to its extensive diversity of topics. Trained librarians should be consulted to make best use of the stock.

At the point of publication the publisher may assign a code number to a book ready for cataloguing. Alternatively, the local librarian (who seldom specialises in instrumentation) decides where a book, or periodical, should be placed. It is often possible for a book to be placed in one of many alternatives – it is not unheard of for a second edition of a book to be located differently to the first!

Libraries generally use either the Dewey Decimal, Library of Congress LC, Universal Decimal UDC classifications, or a locally generated topic system.

The lack of structure of the knowledge of instrument science is reflected in library holdings: material will be found diffused throughout many classes. An impression of the situation is obtained by study of the selected list of subjects and code numbers given in Appendix 3. (The Dewey classification contains over 25 000 named code numbers of which some 800 are instrument related.)

Libraries generally maintain an alphabetical subject index but these will seldom provide the detail sought. The wide choice of instrument topics requires searching based on detailed study of the handbook of the classification system.

The UDC system can provide extreme classification detail. In practice, however, some libraries only classify to the *abridged* version of UDC.

4.2. Classification by literature authors and readers

Turning to classification from the authors and readers viewpoint it is again found that numerous possibilities exist for deciding the key list of index terms.

In general authors align more with their field of application of discipline than with sources publishing fundamentals of instrument science. Often they see the instrument content as insignificant and not worth abstracting. Many facets of a report might be chosen as keys. The reality is that the report, although given several *descriptors* can only be published in *one* journal.

As an example consider an hypothetical report containing, as part only, the 'calibration of a novel, laser-based, alignment system used, by the XYZ Company, to test the military load-carrying capacity of an historic concrete bridge whilst it is conveying a mobile nuclear reactor'. Table 1 gives many likely key features chosen to classify the report. The report is only to be published in one journal – which one would you choose? If the instrument content is minor it may not be abstracted as a contribution to instrument science at all! Authors should give attention to the need to prepare several papers on a major topic, each on the key aspects.

5. Locating book material

Conventional procedures for locating book material in a library system are well described in library guides. In essence the user consults the subject catalogue (usually cards prior to the mid 1970's acquisitions, microfiche subsequently) to locate catalogue locations and specific titles.

Except for the very few librarians trained in instrument literature, and bearing in mind the earlier comments, it is helpful to have carefully selected key words ready to motivate the mind. Book literature is not allocated extensive key words, as are journals, merely a code number. It is helpful to consider, along

---

Table 1. Classes and example journals in which an instrument might be reported (for example, see text)

Class basis chosen	One example of typical journal
Physical principle used (interferometer)	*Applied Optics*
Discipline based on a device used (laser)	*Laser Focus*
Contemporary nature (novel and timely)	*New Scientist*
Military relevance	Confidential report of military organisation
Discipline of use (surveying)	*Photogrammetric Engineering*
Parameter of measurement (measurand is related to strain)	*Journal of Strain Analysis*
Measurement principle used (angle measuring interferometer)	*Journal of Physics E: Scientific Instruments*
Constructional material used (concrete)	*Magazine of Concrete Research*
Standardisation (calibration)	*Newsletter of National Conference of Standards Laboratory*
Design implications	*Journal of Elasticity*
Testing	*International Journal of Non-destructive Testing*
Instrumentation detail	*Instrumentation and Control Systems*
Engineering heritage	*Transactions of the Newcomen Society*
Trade journal	*XYZ Affairs* (hypothetical name)

Plus other aspects such as nuclear systems, transportation, power generation, environmental issues, civil engineering and many more.

the lines of table 1, the many classes into which a work might be placed.

To help people locate book literature of instrument science the IMEKO Higher Education Committee membership produced a significant bibliography, Sydenham (1980a). The most expedient method of compiling the 860 titles content was found, not to be to use computer based methodology, but to obtain data from publisher's catalogues and the personal booklists of experts in instrument science.

Appendix 4 is a checklist of reference sources available in most libraries.

A considerable amount of valuable instrument-related information exists that was published prior to the 1950's, a date about which many libraries discard, or store, holdings as being outdated. This is especially true for mechanical aspects of instrument design which have seldom been published since those times. Recovery of early literature is possible through the bibliographies on mechanical design in Sydenham (1980b) and on the general historical development of instrumentation in Sydenham (1979).

The Instrument Science and Technology series of this journal, of which selected papers are published, Jones (1982) and the handbook, Sydenham (1982) each provide extensive bibliographies on current instrument science. Publications pertinent to instrumentation in the fringe sciences are reviewed in Van Brakel (1977).

6. Locating journal material

Retrieval of journal based information was, prior to c. 1970, generally based upon use of *abstracts* of the content and author details of papers, these being printed in abstracting periodicals. Electronically based data processing led, by the late 1960's, to the information first appearing in electronic format from which the abstracting material was then published in printed form. This provided the opportunity to retrieve literature abstract information by direct EDP methods. During the 1970 decade many computer data bases were created to contain the abstract information and provide the information by sort routines. Users could initially obtain *retrospective* searches by providing a key words list, a service that was produced *off-line* and taking several days to provide readout.

The need to be able to tailor the search pattern, as information was gained by searching, led to *on-line* search operation.

It is now commonplace to use a local computer terminal keyboard to communicate with computer data bases in the user's own country or in other countries. Appendix 5 is a selected list of bases that are likely to be useful for instrument science material. Again the diversity of places where suitable information might be available can lead to the need to search many such bases – with subsequent high cost.

The apparent efficiency and low-cost of on-line retrieval can be misleading. Several shortcomings need to be appreciated.
(i) Few bases contain material published prior to 1970.

(ii) None (bar GIDEP, which is not openly available) is specifically concerned with instrument science: to extract detailed instrumentation information requires a carefully devised search routine (trained operators are mandatory as an aid to users as they appreciate the instructions required when on-line).
(iii) Not all of the total world output of periodical journal literature is included – some is rejected as being obscure or ephemeral. Report and other parochial literature is gradually being placed in EDP systems.
(iv) Only abstract information is immediately available, the user still has to locate and procure the full text. Abstracts can often be most misleading!

7. Keeping up to date

7.1. Current awareness

The vast quantity of information continuously produced makes it impossible to keep up to date in a field by the browsing method alone. This still has its part; regular scanning of selected new periodical receipts acts to bring attention to new developments and news not indexed.

Interaction with workers in the same field, exchanging reprints and attending conferences also play an important part in current awareness.

On a regularly organised basis, information on books is to be obtained by participation in such publisher's services as the International Book Information Service IBIS.

The vast journal literature can be followed by use of services provided by data base proprietors. The publications *Current Contents* and *Current Papers* provide listings of papers just, or even about to be, published.

As well as the printed lists publications, subscribers can purchase a service wherein they receive, on a regular basis, printed abstracts on their chosen area of interest. It is preferable, before commencing this form of service, to first conduct an on-line, retrospective search of the interest using the result to establish keywords and a search profile.

7.2. Personal data bases

Browsing, current awareness services and other sources provide the individual with a steady influx of quite particular information. For this to be useful it, in turn, requires ordered storage.

Traditional personal methods involve card indexes including those that can be manually sorted by indexing holes.

When the number of entries reaches a few hundred (which can occur rapidly) it becomes worthwhile to enter the references into a personal computer or into available centralised services that can accommodate private files. Such computer based options require only elementary sort routines because the problem of locating a few references in a thousand is far easier than from a few million entries.

Acknowledgments

The assistance of librarians at the Skinner Library, The City

University, London and Levels Campus Library of the South Australian Institute of Technology is appreciated.

Appendix 1

Selected standards documents of terms used in metrology

BS 2643:1955 'Glossary of terms relating to performance of measuring instruments'

IEC 50(20):1958: 'Scientific and industrial measuring instruments'

(OIML) PD 6461:1969 'Vocabulary of legal metrology – fundamental terms'

IEC 50(00):1975 'General index of the International Electrotechnical vocabulary'

BS 5233:1975 'Glossary of terms used in metrology'

AS 1514:Pt.1-1980 'Glossary of terms used in metrology – Part 1 General terms and definitions'

IEC-Technical Committees for electrical ᴄ for electronic measuring equipment issued 'Draft-International vocabulary of basic and general terms in metrology' Document$^{13}_{66}$(Secretariat)$^{257}_{44}$ May, 1981.

Appendix 2

Some instrument-related terms from the 1979 INSPEC thesauras (of 8000 terms some 650 are relevant to instrument science).

Automatic testing
Data acquisition
Display instrumentation
Error detection
Geophysical equipment
Human factors
Laboratory apparatus and techniques
Logic testing
Measurement theory
Noise measurement
Nuclear instrumentation
Particle detectors
Patient diagnosis
Picture processing
Probes
Remote sensing
Seismology
Sensory aids
Signal detection
Spectral analysis
Transducers
Water pollution

Appendix 3

Some subjects and codes in Dewey Decimal Classification

Edition 19 (from a list of nominally 800 instrument related topics contained within the whole).

Code Number	Subject
001.534	Perception theory
001.64	Data processing, electronic
121.8	Epistemology – worth and theory of values
363.256	Criminal investigation (laboratories)
371.26	Educational tests and measurements
384.7	Alarm and warning system
389	Metrology and standardisation
511.43	Error theory
522.2	Instruments (astronomy)
529.78	Instruments for measuring time
530.16	Measurement theory
534.4	Measurements, analysis and synthesis of sound
537.5	Electronics
542.3	Measuring apparatus (chemistry)
543.07	Instrumentation (chemistry)
620.1127	Nondestructive testing of materials
621.381043	Measurement (electronic)
621.38417	Measurement and standardisation (radio)
622.15	Geophysical exploration
629.135	Aircraft instrumentation and systems
637.32	Cheese production – quality determinations
658.562	Quality control
677.0287	Textiles – testing and measurement
681	Precision instruments and other devices
774	The Arts – holography

Appendix 4

Check list of library reference sources

Cataloguers index volume for classification used in library (such as Dewey Decimal, Library of Congress, Universal Decimal Classification).

Subject indexes in book and periodical catalogue, in data base manuals, in microfiche of national library holdings.

Technical reference encyclopaedias.

Citation sources, such as, *Science Citation Index* – enables author's work to be followed as others make use of it.

Industry commodity trade directories.

Catalogues, such as *Books in Print*.

Abstracting periodicals of discipline concerned.

Reference volumes such as *Cumulative Book Index* and *British Books in Print*.

Periodicals lists, such as, *World list of scientific periodicals* (60 000 titles).

Appendix 5

Some data bases relevant to instrument science

BA Previews	Biological abstracts, bioresearch
CA Condensates	Chemistry abstracts
CDI	Dissertations
CLAIMS	Several bases on US patents
COMPENDEX	Engineering

ELECOMPS	Elecronic components
ENVIROBIB	Environmental science
GEO ARCHIVE	Geosystems
GIDEP (limited access)	includes several topics – reliability, metrology, failure, experience
INSPEC	Physics, electrical, electronics, computers and control
ISMEC	Mechanical engineering
MEDLARS	Medicine
NTIS	US Government R & D reports
SCISEARCH	Multidisciplinary
WPI	World patents

References

Jones B E 1982 *Instrument Science and Technology* Vol 1 (Bristol: Adam Hilger)

Melton L R A 1978 *An Introductory Guide to Information Sources in Physics* (Bristol: The Institute of Physics)

Sydenham P H 1979 *Measuring Instruments: tools of knowledge and control* (London: Peter Peregrinus)

Sydenham P H 1980a (ed.) *A Working List of Books published on Measurement Science and Technology in the Physical Sciences*. International Measurement Confederation IMEKO (Delft: Applied Physics Dept, Technische Hogeschool Delft)

Sydenham P H 1980b Mechanical design of instruments *Measurement and Control* **13** 365–72 and subsequent issues

Sydenham P H 1982 (ed.) *Handbook of Measurement Science* Vol 1 *Theoretical fundamentals* (1982) Vol 2 *Fundamentals of practice* (in press) (Chichester: John Wiley)

Van Brakel J 1977 *Meten in de Emprische Wetenschappen (Measurement in the Empirical Sciences)* (Utrecht: Workgroup for basics in Physics, University of Utrecht)

Index

147